S0-AIO-815

GEORGE GREEN

The Plaque in Westminster Abbey in honour of George Green.

George Green

Mathematician & Physicist 1793–1841

The Background to His Life and Work

Second Edition

D. M. Cannell

Society for Industrial and Applied Mathematics
Philadelphia

This SIAM edition is a second edition of the work first published by The Athlone Press, London and Atlantic Highlands, New Jersey, 1993.

10 9 8 7 6 5 4 3 2 1

Library of Congress Cataloging-in-Publication Data

Cannell, D. M. (Doris Mary).
George Green : mathematician & physicist, 1793–1841 : the background to his life and work / D. M. Cannell.— 2nd ed.
p.cm.
Includes bibliographical references and index.
ISBN 0-89871-463-X
1. Green, George, 1793–1841. 2. Mathematical physics—Great Britain—History. 3. Mathematics—Great Britain—History. 4. Physicists—Great Britain—Biography. 5. Mathematicians—Great Britain—Biography. I. Title.

QC16.G64 C36 2000
530.15'092—dc21
[B]

00-041938

- The frontispiece appears courtesy of the George Green Memorial Fund.
- Plate 21 appears courtesy of Andrew Dunsmore, Picture Partnership.
- Plate 22 appears courtesy of the University of Nottingham School of Physics and Astronomy.
- Appendices VIa and VIb appear courtesy of the University of Nottingham *Gazette*, Clarice Schwinger, and Freeman J. Dyson.

TO

LAWRIE CHALLIS

Contents

List of Illustrations

Plate 12: A portrait of William Whewell, aged thirty-one, painted by Lonsdale in 1825.

Plate 13: A portrait of Robert Murphy by Dr. John Woodhouse, brother of Robert Woodhouse, painted in 1829, the year Murphy was elected a Fellow of Caius College.

Plate 14: William Hopkins of St. Peter's, a renowned Cambridge coach. He gave William Thomson one of his copies of Green's *Essay* in 1845, prior to Thomson's departure for Paris.

Plate 15: William Thomson, later Lord Kelvin, in 1845, the year in which he discovered Green's *Essay* and in which he was appointed to the Chair of Natural Philosophy at Glasgow University.

Plate 16: The title page of Green's second memoir on light, published in the *Transactions* of the Cambridge Philosophical Society of 1839. The inscription is to his maternal cousin, Lewis Hartwell, who was also one of his executors.

Plate 17: Mrs Jane Moth, the mathematician's eldest daughter, who later became Mrs Stephen Pickernell. This photograph was taken in Richmond, Surrey, when Mrs Moth was possibly in her fifties.

Plate 18: Members of the Mathematical Section of the British Association for the Advancement of Science at the grave of George Green in 1937. H. Gwynedd Green, wearing glasses, is standing to the left of the chairman laying the wreath.

Plate 19: The derelict Green's Mill, before restoration.

Plate 20: Green's Mill in 1985, restored to full working order.

Plate 21: The Dean of Westminster dedicating the plaque to George Green; behind him Sir Michael Atiyah, Professor Lawrie Challis and Mary Cannell are seated below the memorial to Isaac Newton.

Plate 22: (From left to right) Professor Freeman Dyson, Professor Lawrie Challis, and Professor Julian Schwinger in conversation following the public lectures on Green on July 14, 1993.

Preface to the Second Edition

There had always been the hope that the biography of George Green, originally published in 1993, would at some time be reprinted, since this would make the story of Green more readily available to a younger generation of scientists. There are few who do not come across Green's functions and his Theorem in the course of their studies and some will continue to use them throughout their careers.

Recently, it has become evident that there is increasing interest in the history of mathematics and in science as a whole. In addition to research in the development of ideas and concepts, there is a parallel curiosity about the men who produced them. As in the case of Green, some were the victim of circumstances; their work and their lives were shaped by various factors in the period in which they lived, their social status, the people they met and the degree of understanding with which their findings were received. Each individual case adds to the general picture; recent research has brought George Green more clearly into focus.

My thanks are due to SIAM, and in particular to Professor Robert E. O'Malley Jr., for making the present reprint possible. One deficiency in the original publication was the sparse information concerning contemporary awareness of the nature of Green's wider contribution to modern physics and technology. To some extent this has been remedied by the addition of a new appendix and an expanded list of scientific references. The last chapter of the Green narrative now includes an account of the Bicentenary Celebrations of his birth in 1993, only in the planning stage at the time of the original publication. These included two major events: national recognition of George Green through the dedication of a plaque in his honour in Westminster Abbey in London, and international recognition through public lectures given in the University of Nottingham. These were delivered by two of the leading scientists in quantum electrodynamics of this century, Julian Schwinger,

Nobel Laureate, and Freeman Dyson, F.R.S., of the Institute of Advanced Study, Princeton University, Princeton, New Jersey. Schwinger would seem to have been the pioneer in giving Green's functions a fresh and vigorous life in quantum physics. His lecture, along with Freeman Dyson's account of his earlier work in quantum field theory, throws light on a recent period of research of interest to younger mathematicians and scientists. Discussions of Green's importance in modern physics and technology are to be found in notes of lectures given during the Bicentennial Commemoration at the Royal Society in London (see Note 22 to Chapter 11).

Major acknowledgements are recorded with gratitude to Mrs Clarice Schwinger, who generously agreed to the inclusion of her late husband's paper on Green, and to Professor Freeman Dyson, who equally generously agreed to my including his paper. My debt to those who earlier gave permission for the inclusion of material in the original publication still stands high. Indeed a retrospective view of its reception would indicate a strong interest in the circumstances of Green's life. This confirms the importance of the two main sources which supplied the essential information. The first is the original research by H. Gwynedd Green into the life of the mathematician and his family in Nottingham, and I would like to reiterate my thanks to his daughter, Mrs Hazel Bacon, for the use of this essential material. The unique glimpse (in the absence of other material) into George Green's mind and character is found in his letters to his patron, Sir Edward Bromhead, and I am grateful to his descendant, Mrs Anne German, for continuing to provide access to the family papers. For a non-mathematician, the help of a former collaborator, N.J. Lord, has proved invaluable, particularly in providing the notes on lectures given at the Royal Society Commemoration. Since 1993, the considerable importance of Appendix I by M.C. Thornley has become apparent. It takes readers to the coal-face, as it were, by introducing and elucidating the actual text of Green's famous and seminal Essay of 1828. I would like again to express my gratitude to him for providing such a significant contribution to our presentation of Green.

Professor Lawrie Challis, as in the past, has continued to give his stalwart support to this further development in the fortunes of George Green. Despite manifold scientific and professional pressures, he has always been ready, with his customary tact and wisdom, to offer help, share problems, and give advice when requested. I would also like to thank Mrs Jennifer Challis who, of similar mind and spirit, was most helpful at a particularly crucial time.

As an adopted member of the Nottingham University Physics Department, (now the School of Physics and Astronomy), I wish to express appreciation of much friendliness, interest and valued help; to Professor Colin Bates, Head of the Department and his colleagues; to Mrs Linda Wightman, for her efficiency, patience and good humour in acting as an unofficial secretary over the Green years; and to Mr Terry Davies, Department photographer, who provided additional photographs, including the highly prized one, taken in less than ideal circumstances, of Professors Schwinger, Dyson and Challis in conversation in the Physics Department after the public lectures. The equally prized lectures in the present volume are reprinted from the Nottingham University *Gazette* by kind permission of the University. I would also like to express my grateful thanks to Ms Deborah Poulson, Developmental Editor at SIAM, for her guidance in preparing the reprint for publication; her patience and quick response to queries have been much appreciated.

Finally I must record deep gratitude to a recent collaborator, Mr John Clayton, whom I had the good fortune to meet a few years ago. With his professional experience and expertise, he has combined the services of word-processor operator and copy editor; without his generous help, this reprint would not have been possible.

Preface

This is a story written to interest the lay reader as much as the scientific specialist. Detailed discussion of Green's published papers has therefore been avoided, with the exception of the Essay of 1828, the main points of which are discussed in Appendix I.

The facts relating to George Green and the history of his family have already been detailed in two publications. The first was the monograph of 1945 by the late H. Gwynedd Green, who was the pioneer in researching the life of George Green. As a graduate of Caius College Cambridge he developed a lifelong interest in Green, which he pursued during the thirty-three years he spent in the Mathematics Department of Nottingham University, and until his death in 1977. I am very grateful to H.G. Green's daughter, Mrs H.M. Bacon, for permission to make unrestricted use of her father's work.

The second account of the Green family may be found in the middle section of the Nottingham Museum booklet *George Green, Miller, Snienton*, published in 1976. This was written by the then Senior Archivist of the Nottinghamshire Record Office, Mrs Freda Wilkins-Jones, who was able to trace a more detailed picture of the family history from its Saxondale origins through to its apparent extinction this century. I am glad to acknowledge her important contribution to the history of the Green family and the additional information she kindly allowed me to use in my previous publication. The account of Green's relations with Sir Edward Bromhead would not have been possible without reference to the correspondence originally discovered by Dr J.M. Rollett and painstakingly transcribed and published in the last section of the Castle booklet by its editor, Mr David Phillips, then Keeper of Art at the Castle Museum. I am indeed grateful to Mr Phillips for his generous and spontaneous agreement to my use of this material, as

I am for the guidance given in his account when I retraced his steps on visits to Gonville and Caius College Cambridge.

As these two publications were out of print, it was thought desirable to republish this valuable material, and incorporate it into a narrative which would also aim to fill in some of the background to episodes in George Green's life, in a form which would appeal to the general public. The time was appropriate, since the restoration of Green's Mill has brought many visitors to Nottingham, curious to know more of the life of the mathematician. There was also the fact that the bicentenary of George Green's birth is due to be celebrated in 1993.

In researching Green's background, I was given invaluable help by a number of people. One of the first of many kind friends was Dr J.M. Rollett who, having previously located the Bromhead correspondence, readily and generously shared his findings on Bromhead and Cambridge, and later his genealogical researches on the Green descendants. Mrs Anne German (*née* Bromhead) offered a warm welcome on my visits to Thurlby to discuss the history of her family. I am indebted to her and to her husband, Mr Robin German, and also to Mr James Hall and his sister, Mrs Elizabeth Draper, descendants of George Green Moth's first marriage, for permission to reproduce the Green–Bromhead correspondence, the silhouette portrait of Edward Bromhead and the photographs relating to the Green family. Copies of these documents have been deposited in the Department of Manuscripts and Special Collections, Hallward Library, University of Nottingham. I owe much to the University Librarian, Mr Peter Hoare, and his colleagues, in particular Dr Dorothy Johnston, Mr Neville Green and Mr Michael Brook, for their unfailing help and resource. Mrs Joan Taylor and Mrs Jane Corbett, her successor as Librarian of the Nottingham Subscription Library at Bromley House, provided ready access to the Library's minutes and records. I am likewise grateful to Mr Adrian Henstock, County Archivist, and his Deputy, Mr Chris Weir, at the Nottinghamshire County Record Office; and to Mr Stephen Best, until recently at the Nottingham County Library, especially for his detailed knowledge of Sneinton.

Some of the most rewarding hours working on this biography have been spent in Cambridge. Naturally, Green's college, Gonville and Caius, provided much essential information, and I must record my gratitude to those who welcomed me there and offered me their knowledge and friendship: to Professor Christopher Brooke, historian of the College, who guided me through the chapters on Cambridge, and to Dr Anthony Edwards, fervent admirer of both Green and Bromhead; to Mrs Catherine Hall, College Archivist, and her successor, Ms Anne Neary; and to Mrs Alison Sproston, College Librarian, and her predecessor, Miss Heather Owen. From Caius, the Green trail took me to Queens' College, where the Curator of the Old Library, Mr Iain Wright, and Mrs Clare Sergent, shared with me the excitement of investigating Isaac Milner's inventories, and Dr John Twigg, College historian, provided valuable information on John Toplis. Miss Anne Stow, Secretary of the Cambridge Philosophical Society, produced minute books and records which allowed me to trace Green's connections with the Society, and in particular to investigate the intriguing discovery of the Jacobi papers. Mr Nicholas Smith, Librarian in the Rare Books Reading Room at the University Library, and Mr Michael Petty, at Cambridge County Library, each provided most useful material on Cambridge booksellers.

The project of writing this memoir would not have materialized without the loyal support of friends who are interested in making the name of George Green more widely known. From the beginning, Professor Lawrie Challis has been constant in his encouragement and practical help. It has proved both a privilege and a pleasure to support him and his colleagues in their project of establishing wider recognition in the scientific community and in the public mind at large for the genius of George Green. In particular I am most grateful to Professor Challis for reading the script and writing the Foreword.

Full appreciation of the scope and depth of Green's work is arguably possible only for the mathematician and physicist working in reasonably advanced areas of their discipline. It was felt, however, that some indication of its character should be provided. This has been done with clarity and conciseness by Mr Michael Thornley who, despite other claims on his time, produced Appendix I, which contains an analysis of Green's Essay of 1828 and an interpretation of some of its more archaic features for the benefit of modern scientific readers. I am most grateful to him for an indispensable contribution to this biography. Dr John Roche, of Linacre College Oxford, gave long-term support and encouragement; his suggestions and much-needed critical appraisal of the script are gratefully acknowledged. A collaborator of long standing is Mr Laurence Tate, who, with his daughters Helena and Barbara, typed and processed numerous drafts of the text with limitless patience and understanding. The production of this book would not have been possible without their expertise and sustained interest. Neither would it have been possible without the detailed help and advice I received from the Chairman of The Athlone Press, Mr Brian Southam, and his colleagues, Mrs Gillian Beaumont and Ms Helen Drake, who guided a first-time author through the unfamiliar paths of professional publication. Mr David Williams most kindly gave his time to read through the proof sheets of the text. I would like to acknowledge the co-operation of Chas. Goater and Son in providing copies of two illustrations from my previous publication. Most generous support in the publication of this biography has been given by the Physics Department of Nottingham University, the George Green Memorial Fund and the Barbara Dalton Trust of the Sneinton Environmental Society. Lastly, I would like to thank my family and my many friends and correspondents who, over a number of years, never failed in their interest and encouragement, particularly when it was most needed!

It will be evident, both from these acknowledgements, and from the Notes to the text, that the writer of this narrative has functioned more as editor than as prime researcher. In bringing together from many different sources the various pieces of information which constitute the life of George Green, I may perhaps, in common with others, claim with Montaigne: 'I have gathered a posy of other men's flowers and nothing but the thread that binds them is my own.'

Acknowledgements

In addition to my expressions of appreciation in the Preface, I would like to record my thanks to the Master and Fellows of Gonville and Caius College Cambridge, who kindly allowed the portrait of Robert Murphy to be photographed (by Ed. Collinson of Cambridge) and reproduced in this book; to the Master and Fellows of Trinity College Cambridge for permission to quote from the letters of William Whewell, and to reproduce his portrait and those of Charles Babbage and John Frederick Herschel; to the President and Fellows of Queens' College Cambridge for permission to reproduce the portrait of Isaac Milner; to the Master and Fellows of Peterhouse Cambridge for their kind permission to reproduce the portrait of William Hopkins; to the Cambridge Philosophical Society for permission to reproduce the portrait of George Peacock; to the British Library for permission to quote from Sir Edward Bromhead's letters to Charles Babbage; to the Librarian, Glasgow University Library, for permission to quote from William Thomson's unpublished letter to George Boole, and to reproduce his photograph; and to the President and Council of the Nottingham Subscription Library in Bromley House, Angel Row, Nottingham, for permission to reproduce the photograph of Bromley House. Despite enquiry, I have been unable to trace the photographer of the excellent photograph of Green's Mill, but would like to make due acknowledgement here. For the use of the other illustrations I am indebted to the Manuscripts Department of Nottingham University for the use of nos 2, 3, 4, 16 and 18; and Nottingham City Council, Leisure and Community Services Department, for no. 19.

Foreword

by Professor Lawrie Challis

George Green was a pioneer in the application of mathematics to physical problems. Latimer Clark, a nineteenth-century historian of electricity, described his Essay of 1828 as 'one of the most important works ever written on electricity'; and Sir Edmund Whittaker, in his authoritative *History of the Theories of Aether and Electricity*, states that 'it is no exaggeration to describe George Green as the real founder of that "Cambridge school" of natural philosophers of whom Kelvin, Stokes, Lord Rayleigh and Clerk Maxwell were the most illustrious members in the latter half of the nineteenth century'. The significance of his work on solids is described in Love's *Mathematical Theory of Elasticity* as 'the revolution which Green effected in the elements of the theory'.

Clark, Whittaker and Love were writing of the influence of Green's work on classical physics, but the scale of its application to modern physics is perhaps even more remarkable. It started in 1948 when Julian Schwinger, who had been using Green's functions in his wartime work on radar, used them for the first time to solve a problem in quantum mechanics. The problem was on quantum electrodynamics – how light interacts with electrons – and Schwinger shared a Nobel Prize with Feynman and Tomonaga for solving it. This pioneering step was followed by a dramatic increase in the use of Green's functions, particularly in the theory of fundamental particles and nuclei, and in solid state physics. An assessment of the value of the technique is given in a letter from Robert Schrieffer, who shared the Nobel Prize for the theory of superconductivity with Bardeen and Cooper:

> I have, in most of my scientific publications, dealt in one way or another with the techniques of Green functions. . . . Not only are Green functions of great significance to the theoretical

physicist in the solution of physical problems, these functions are directly related to physical observations in the laboratory. Almost every experiment which weakly probes a physical system can be described in terms of the relevant Green's function for this observation. Thus the theoretical physicist has a direct link to the experimental results through the work of George Green.

But while practising scientists had – and have – no doubt of Green's standing as a mathematician and physicist, his name is much less known outside the scientific community than those of other nineteenth-century British scientists such as Faraday, Joule, Kelvin and Maxwell. Part of the reason for this must surely be related to the advanced nature of Green's work. All of it lies beyond the school curriculum, and while this is also true of much of Kelvin's work, his name is familiar because of the Kelvin scale of absolute temperature.

In this book, the first major biography of George Green, Mary Cannell describes the life and background of this remarkable man. There were earlier articles which provide some of the material, starting in 1850 with Kelvin's preface to Green's Essay reprinted in *Crelle's Journal*, and she has been generous in acknowledging the debt she owes their authors. Her book, however, gives by far the most complete picture of Green's life and education in both Nottingham and Cambridge, and of the people who had an influence on his creative development. She has also given us insight into the nature of the relationships that would have existed between Green, a miller in a provincial town, and the fringes of the scientific establishment with which he made cautious contact. It is a fascinating book.

I first met Mary Cannell in 1977, soon after the start of a campaign to restore Green's Mill. She had recently retired as Acting Principal of Nottingham College of Education, then one of the largest Colleges of Education in England, and I was very grateful when she agreed to become Honorary Secretary of the George Green Memorial Fund. At that stage I had thought only of harnessing to our cause her very considerable energy, determination and organizational ability; I had not dreamed that she would also spend much of the next fifteen years unravelling Green's story. The work has taken her around the country talking to Green's descendants, many times to Cambridge, particularly

Caius and Queens' colleges, to Edinburgh and Manchester and even to Melbourne, Australia, to talk to J.J. Cross, an authority on Green's period in the history of mathematics. She has shared her enthusiasm with many; she is a vivid lecturer and has given well over fifty lectures on Green. Her book will allow this enthusiasm to reach a much wider audience, and we are greatly in her debt. Her claim that she has worked more as editor than as prime researcher is far too modest. She has illuminated our picture of Green in many ways, and one that I find particularly interesting is her identification of Toplis as the Nottingham mathematician most likely to have guided Green's education after his eighteen months at school. I have found only one omission in the book, and that is reference to the vital role she herself played in the restoration of Green's Mill and the organization of his bicentenary celebration. Through this work on Green, Mary Cannell has made many friends in the local and scientific communities, and I am delighted to have this opportunity of thanking her on their behalf and mine for her work, and for this intriguing book.

Professor Lawrie Challis
Nottingham University

Introduction

There are few mathematicians and physicists working in the advanced stages of their specialism who have not at some stage encountered Green's functions, yet not many of them know anything of their eponymous creator. The name of George Green, who devised this powerful and widely used technique, is not always found in earlier dictionaries of science, and more recent publications retail the same few facts and present sometimes an inadequate account of his achievements.

The present work sets out to investigate the circumstances in which Green found himself, to place the known facts in their context, and to trace the different stages of his life in which he produced his various investigations. This may go some way to explain how a man who was perhaps the first mathematical physicist in England came to be so little known as an individual to the generality of scientists and completely unknown to the public at large. It is the aim of this biographical memoir to make available to that public what is known of George Green, and to present him as a person rather than merely the inventor of a mathematical function. Green's biographer has a more difficult task than the biographer of Kelvin or Faraday, for example, since Green, unlike them, left little in the way of correspondence, and no diaries or working papers. Any that may have survived were destroyed, it would seem, early this century. The biographer of George Green is left, therefore, with the facts culled from official sources, a dozen letters, and a few personal statements from a handful of individuals; neither is there a likeness of him, either portrait or photograph. Yet in the light of Green's achievements, a record of his life should surely be attempted.

George Green, born in 1793, was the only son of a Nottingham baker. George Green Senior, semi-literate but with a head for business, prospered sufficiently to build a new windmill at Sneinton, a village just outside the town boundary, and grind his own corn for the bakery. Some years later he built a substantial family house next to the mill, where George Green and his parents went to live. George Green was a reluctant miller, but he had no choice in the matter until his father died in 1829, and he found himself sufficiently affluent to be able to dispose of the milling business and devote himself to mathematics. This had been his passion since his youth – so much so that his father had sent him at the age of eight to the town's leading academy, run by Robert Goodacre, an enthusiast for 'the mathematics' and for the popular science of the day. Young George stayed only four terms, by which time he had learnt all his masters could teach him, so he was set to work in his father's bakery. From then until 1823, over twenty years later, there is little information on Green, and none on his personal development.

In that year, George Green joined the Nottingham Subscription Library. There he had access to books, and also to journals and periodicals. Equally important was the fact that membership of the Library eased the thirty-year-old miller out of his intellectual isolation and the company of manual workers, of farmers and shopkeepers, millers and stablemen, into a different stratum of society: that of gentlemen of leisure, of professional men such as clergymen and doctors, and well-to-do businessmen. For them the Library, situated in Bromley House in the centre of Nottingham, functioned as a gentlemen's club, where politics and public affairs were debated and current topics of social and scientific interest were discussed.

Association with these people and the use of the Library facilities were beneficial for Green. In five years he had written his first work, 'An Essay on the Application of Mathematical Analysis to the Theories of Electricity and Magnetism'. This was published in 1828 by private subscription and more than half the fifty-odd subscribers were members of Bromley House. The Essay attracted scant attention, however, and despondently, Green returned to his milling, until in 1830 he came into contact with Sir Edward Bromhead, a Lincolnshire landowner and mathematics graduate of Gonville and Caius College Cambridge. Bromhead recognized the originality of Green's work and encouraged him to resume his mathematical studies. As a result, Green wrote three papers which Bromhead sponsored for publication in the *Transactions* of the Cambridge Philosophical Society and the Royal Society of Edinburgh. Finally, in 1833, Green, at the age of forty, enrolled as an undergraduate in Bromhead's own College, Caius. Green took the Mathematical Tripos in 1837, emerging as Fourth Wrangler, and two years later he was elected into a College Fellowship. During these two years he published six more papers in the *Transactions* of the Cambridge Philosophical Society, of which he became a member. Ill-health compelled him to leave Cambridge some two terms after his election. He returned to Nottingham, where he died in May 1841 at the age of forty-seven. In his early twenties, Green formed a relationship with the daughter of his father's mill manager, Jane Smith. She bore him seven children, the last just thirteen months before his death. They never married, though Jane and the children were known by the name of Green.

Green's best known work is the Essay of 1828 in which he introduced the term potential in electricity and, more importantly, Green's functions and Green's Theorem. In his Cambridge papers, Green abandoned the topics of electricity and magnetism and made original contributions in hydrodynamics, sound and light.

Such are the bare facts of George Green's life, but they provide no answers to the many questions that inevitably arise. It is normally claimed, for example, that Green was self-taught, and to a considerable extent this is true, given the nature of his achievements. It is also true that self-education was a feature of the period, and a number of those mentioned in this narrative – Isaac Milner, George Boole and Michael Faraday – had no formal education such as Babbage, Bromhead or Kelvin received. One is entitled to question, however, whether there was anyone who set Green on his mathematical path, and the enquiry is the more pertinent since Green's mathematical knowledge was of a very specialized kind. How did he acquire his knowledge of the 'Mathematical Analysis' of the Essay title? Green showed himself to be fully at home with – indeed, he improved upon – the system of mathematics used on the Continent, at a time when it was almost completely unknown in England and ignored in Cambridge, which was then the main centre of mathematical studies in the country. Yet here was a working miller in a provincial town, whose only documented contacts with mathematicians in Nottingham were one or two graduates in Bromley House Library who had graduated in Newtonian mathematics many years before. A further question is what led Green to the subjects of electricity and magnetism at a time when these were not current topics of interest for men of science? His sources in the Essay were papers by Poisson, published in Paris between 1811 and 1825, but these were certainly not on ready sale in the bookshops of Nottingham.

Green's personal life also presents mysteries. Some discretion has been exercised in the past on the subject of his family life, which has caused dismay or disappointment, if not disapproval, on the part of his admirers. Why did he not marry Jane Smith and legitimize his children? Given the ultimate recognition of his contribution to mathematical science, why has he remained totally unknown to the townspeople of Nottingham, where he lived for forty years? Various clues to these puzzles have been followed up in an attempt to find some answers. Gradually, something of George Green the man appears – still largely a silhouette rather than a picture in full colour, but firmer in outline than the previous faint impression. Here and there are odd clues which suggest his temperament. But in all cases in this narrative, in the absence of firm evidence, the reader is left to form an independent judgement, and to share in the excitement of tracing the lineaments of this extraordinary man.

A tragic aspect of Green's life is the fact that he was not born at a time or in circumstances conducive to the full recognition of his intellectual originality. Unfortunately for Green, mathematical development in England was at a low ebb in the early decades of the nineteenth century, with Cambridge stagnating in the shadow of Newton, who had produced his mathematics nearly a century and a half earlier. This dead hand of tradition, which stifled much initiative and originality, was in sharp contrast to the situation in France, where a man such as Green might well have been nurtured and reached his full potential. There, in the fifty years from about 1790 to 1840, there was enormous progress in mathematics and in mathematical physics, culminating in the emergence of what came to be known as mathematical analysis. There were researches and discoveries in the fields of heat, electricity, magnetism, optics and chemistry; there was widespread discussion and collaboration between individuals and societies, and the dissemination of findings through papers, books and journals; there was the promotion of organizations devoted to the application of scientific knowledge to practical purposes, and to the education of the next generation of scientists. Finally, in post-revolutionary France, entry into this activity was solely on merit: poverty, social status or religious creed was no barrier. But Green was born in England, not France. Given the situation in Cambridge, it is just as well he did not go there until his mathematical thought had matured. As an undergraduate of normal age, his originality might well have been stifled and he might never have achieved what he did. As it was, his papers of 1828 to 1839 were not published at a propitious time, and fell largely on infertile ground.

If there was lamentably little activity in Cambridge, there was, on the contrary, a drastic change in Green's home town, Nottingham. The pleasant, spacious garden town his father had known when he first set up his bakery in the 1780s became, in the next fifty years, one of the worst towns in England, condemned in the Report of the Royal Commission of 1845 for its slums, its enclosed courts of back-to-back houses, and its high infant mortality. Green Senior built some of this property, as did many others. These years marked the start of the Industrial Revolution, and the town population doubled in half a century. In Nottingham, the beginning of capitalism saw the development of small textile factories, where the installation of more efficient machinery threatened the livelihood of hundreds of near-starving workers, ever ready to rise against those who they felt were responsible for their misery. George Green was a child when his father's bakery was attacked during the Bread Riot of 1800; he was a grown man when he defended his mill against the mob during the Reform Riot of 1831. But the Green family prospered as a result of these changing conditions. George Green Senior, a poorly educated farmer's son, through hard work, tenacity of purpose, a good business head and independent spirit, died a man of property, even able to describe himself as 'gent.' in later years, and left sufficient wealth for his son to live on his income thereafter.

Economic prosperity brought social advancement, and George Green, now in his thirties, was able to rise from the class of self-employed manual worker-cum-tradesman to which one might ascribe an independent miller and mill-owner to that of middle-class gentleman and academic. How easily he adapted personally to this change is not clear. Reports suggest that he was not much involved in Bromley House society, and his early dealings with Bromhead are respectful, to say the least, though it is evident that later he felt more at ease. Green's first report from Cambridge is cheerful, but there is no indication of the nature of later contacts in the University. One senses that he would have resented empty patronage: his natural reserve and the independent spirit of his *concitoyens* of Nottingham would have discouraged that. Bromhead's patronage was practical and of a different order, though his private view of the relationship, as revealed in his letter of 1845, gives reason to pause for thought. It is worth remembering that George Green lived at a time when social distinctions were strong. They were a natural fact of life, and should be recognized as such by a modern reader without the overtones of snobbery they would suggest today. In this historical account, it is important to view such distinctions with the eyes of Green and Bromhead, by whom they were accepted.

Green lived at the time when Georgian England was giving way to the new Victorian England which would become powerfully established during the nineteenth century. He stands between two men of very different social backgrounds – the one a last representative of a disintegrating social structure, the other an example of the new Victorian age, yet all three were born within six years of each other. Edward Bromhead was a country gentleman, born to property and privilege, a landowner and magistrate involved in national and local affairs, a dispenser of patronage and charity, an eighteenth century amateur mathematician, antiquarian and botanist. He contrasts strongly with William Tomlin – who like his cousin, the mathematician – came from farming stock, but swiftly set himself up as a man of affairs in Nottingham, early claiming gentleman status, buying a good town house, and embarking on property and financial transactions. Tomlin was a self-made man, an early example of the new Victorian, acquiring his wealth through personal acumen and a flair for profitable investment. Though his cousin George Green never travelled faster than a galloping horse, William Tomlin lived to make a fortune in railways.

The changing social conditions of the early nineteenth century brought about a change in moral attitudes. George Green, begetting his children in the circumstances then prevailing, was in all probability following a more liberal practice of the previous century, but he left them the victims of Victorian prejudice and intolerance. The standards of nineteenth-century England were formed as a result of the efforts of Nonconformist and other reformers to improve the lot of the poor. Chapels, schools, temperance and friendly societies were all provided or promoted to this end. Unfortunately, the fulfilment of these objectives later induced attitudes of self-righteousness, and even intolerance and condemnation – a charge from which, in modern eyes, even the charitable Misses Tomlin may not be immune. It is open to conjecture whether the suicide of the mathematician's son was in any way due to his illegitimacy; it certainly contributed to his daughter Clara's loneliness and eccentricity. As a result of Green's irregular family life, some at least of his children suffered personal embarrassment, and he personally established no reputation in his home town. His academic achievements were little known outside Cambridge, and he was possibly viewed by his friends in Nottingham as a failed academic – an object of pity, maybe, rather than admiration.

This brief review of the academic and social conditions in which Green lived may go some way to indicate why he is unknown apart from his works. Green received due recognition after his death, however, from those who studied and developed his contribution to mathematical science. The Essay of 1828, which was his first and major work, was unknown in both Cambridge and Nottingham when he died in 1841, and might well have been forgotten. The Essay, however, was crucial to Green's reputation, since it contained some of his most important findings. Through a fortunate sequence of events it was rediscovered by the young William Thomson in 1845. He was instrumental in introducing it to continental mathematicians, and in its publication in *Crelle's Journal* in the 1850s, but the republication of Green's Essay in England, together with his other papers, had to wait until 1871, when it was edited by Ferrers and published in Cambridge by Caius College. Thomson did more than circulate amongst contemporaries the contents of the Essay – which, however, were a stimulus for his own work on electromagnetism. He extolled the primacy and originality of Green's contributions to mathematical physics in general and established his reputation in nineteenth-century science in England. Green's work proved valuable to Stokes and Clerk Maxwell, and to a later generation of scientists such as Love, Lamb, Larmor and Rayleigh. With mention of these last, Green enters the twentieth century. His work in electricity and magnetism has been superseded, but his work in hydrodynamics, sound and light has proved unexpectedly fruitful, though in forms Green himself could never have imagined. The most dramatic resurgence has been in quantum mechanics, when Schwinger and others initiated an apt and extremely successful application of Green's functions. As a result, the name Green is known to twentieth-century scientists worldwide. Indeed, it has recently been established in extra-terrestrial regions. There is a crater on the far side of the moon which was named after George Green in 1970.

This memoir brings together all that is known to date about the personal life of George Green. The search for Green the man started at the beginning of this century and produced Kelvin's letter of 1907 – turning up the first stone as it were. The early Nottingham enquirers, Granger, Becket and their associates, turned up others. In mid century H.G. Green made the first major contribution, which was amplified in 1976 by the publication of further material by a Nottingham group headed by Wilkins-Jones, Phillips, Challis, Bowley and Sheard. In the early 1970s, Professor Challis and colleagues set up the George Green Memorial Fund with the intention of restoring Green's Mill as a memorial to the mathematician. Other university enthusiasts, John Farina and David Edge, contributed articles to journals. Also in the 1970s, J.M. Rollet discovered the Bromhead correspondence and traced the Green family descendants. 1989 produced evidence of a contemporary international contact with the notice of some of Green's papers inscribed by him to Jacobi. A year later came the discovery of Clara Green's grave, and of the later fortunes of the Tomlins. The quest for Green provides continual interest and surprise. What further evidence remains to be discovered? Copies of his Essay perhaps, now described as 'extremely rare'? Further dedications of his papers? More correspondence, revealing unsuspected relationships? Known contacts have been largely exhausted; any further discoveries lie at the feet of Dame Fortune herself.

Mary Cannell
In Memoriam

Mary Cannell died on 18 April 2000 before this second edition of her book was published. She was 86 years old and full of energy and had been working on matters relating to Green until a few days before her death. Indeed, Green had occupied the larger part of her time since her retirement in 1975.

Mary had spent her working life teaching French language and literature in schools and later lecturing in a college of education. Anyone who had the pleasure of hearing her lecture or even of talking to her would have felt the enthusiasm she had for her subject and the authority with which she spoke about it. She must have been an inspiring teacher and role model. She came to Nottingham in 1960 as Deputy Principal of the newly founded Nottingham College of Education in Clifton and in 1974, the start of her last year in the College, she was appointed Acting Principal and had the complex task of overseeing the amalgamation of the College with Trent Polytechnic, later to become Nottingham Trent University. She did this with great sensitivity and skill and on her retirement, the Polytechnic honoured her by making her a Fellow.

The George Green Memorial Fund had been set up, a year or two before Mary retired, with the aim of restoring his ruined windmill, and Mary became an enthusiastic member of a team that worked closely with the Sneinton Environmental Society and the City authorities. She also made a major contribution to the organisation of the 1993 Bicentenary celebrations of Green's birth that involved events in Nottingham, Cambridge and

London, and she gave much input on the design of the Memorial Services at St Stephen's, Sneinton and at Westminster Abbey. Many other activities supported by the fund, such as the naming of a Green room in the Bromley House Library, establishment of the Green website, the location of items for the Green Collection in the University of Nottingham library and the restoration of Toplis' tomb in South Walsham, had been her idea and owed a lot to her in their execution. The Green family became her friends and she wrote a letter to them every Christmas to explain what the Fund had done in the past year. She did all this with tremendous energy and enthusiasm and always with great friendliness and humour.

Mary gave well over 50 lectures on Green to local societies and wrote an illustrated booklet that has so far sold over 5000 copies. The interest this engendered and her realisation that there was much to be done to place Green's life in its context within English social history at a time of rigid class divisions led her to begin work on the first edition of the biography. Her research was extensive and earned her great respect as an historian of mathematics. She was invited to lecture on Green at a number of Universities in Britain, Canada and the United States and her splendid lecture at the Royal Society during the Bicentenary meeting will be remembered for many years. She was made a Fellow of the University of Nottingham in 1995 and an honorary graduate of the Open University and Nottingham Trent University (posthumously) in 1996 and 2000, respectively.

Mary was a very special person who made friends and gained respect wherever she went. She had great warmth and a style and charm that seemed to come from a slightly earlier age. She is greatly missed by her many friends and colleagues.

Lawrie Challis
Nottingham
July 2000

Chapter 1

Family Background

Some six miles along the road east out of Nottingham to Grantham, one comes to the old Roman Foss Way which leads from Leicester north to Newark and Lincoln. The River Trent flows nearby through a green country of fields and farms under a wide expanse of sky. Only the Malkin Hills rise gently a few hundred feet, crowned here and there with a coppice of trees. Shelford lies at their foot, a riverine village with a church, a tavern – The Chesterfield Arms – perhaps a hundred households, and some scattered farms. Saxondale, a mile further up the slope near the main road, is nowadays an even smaller village, with only a couple of farms and a few houses.

This is the country of the Green family, since George Green came from farming stock. For generations his family had farmed land at Saxondale which, since its church had been destroyed at the Reformation,[1] had become part of the parish of Shelford, and it is there that Green's forebears are buried. Nowadays one has to search diligently to find the gravestones lining the edge of the cleared and grassy churchyard. They can be found bordering the far east side near the road, but they are now totally overgrown with ivy and elder. Here are five gravestones of the Green family.[2] The earliest bears the epitaph of John Green, 1651–1721. A later John Green, possibly a grandson, married Ann Girton, who was buried here in 1773. These were the parents of George Green, the mathematician's father. At a time of rural illiteracy, when bride and groom made their mark in the wedding register witnessed by the parish clerk, it is interesting to note that John Green, a tenant farmer, and his son, both signed their marriage licences.[3] The Green family owned five acres of land, but John Green leased 140 acres from the Earl of Chesterfield, as recorded in the Estate Book of 1775.[4] His wife's father, George Girton, leased 191 acres.

Neighbouring tenants were the Tomlins of Shelford. In 1775 Sarah Tomlin leased 124 acres on the nearby Malkin Hills, though in that year control passed to George, presumably her son. The Tomlins and the Greens would later be joined in marriage. The Green family house no longer stands, but it was sited near the present Lodge Farm at Saxondale, where there is still a pump bearing the date 1789 and a large Tudor barn which would have formed part of the Greens' holding.[5]

John and Ann Green had three sons: John, Robert and George. John and Ann died in 1773, within a few weeks of each other. The eldest son, John, then aged nineteen, continued to farm the Saxondale land before moving later to Thurgaton and East Bridgford,[6] leaving his brother Robert to carry on with the family farm. The orphaned George was just fifteen, and as the family farm was presumably not large enough to support three sons, he was apprenticed to a baker in Nottingham, to whom the family paid the handsome apprentice fee of £10.[7] This may have been arranged with the help of William Butler, also a baker in Nottingham. He and his wife both came from Radcliffe-on-Trent, a village only two miles from Shelford and Saxondale and the two families probably knew each other. The Butlers would also be joined to the Greens by marriage.

Nottingham, although it was only six miles away and less than half an hour's travel in modern terms, probably meant a lengthy journey on foot or in a farmer's cart for the young apprentice. Chapman's 1774 map of Nottinghamshire indicates the routes he might have taken: either crossing the Trent by ferry from Shelford to Stoke Bardolph and then on to Nottingham through the villages of Carlton and Sneinton, or along the south bank of the river, skirting Holme Pierrepont and West Bridgford, then crossing over the old Trent Bridge and joining the road into the town from London and the south.

Young George would have found the Nottingham of the 1770s a pleasant, open town with wide streets, stately houses with gardens, and flowered meadows leading down to the Trent. In 1697, Celia Fiennes found Nottingham: 'the neatest town I have ever seen, built of stone and delicate large and long streets much like London',[8] and Robert Sanders declared: 'the situation is not exceeded by any in England. . . . The streets are broad and open and well paved . . .'.[9] Before George Green Senior died in 1829, however, he was to see Nottingham become one of the most overcrowded and unhealthy towns in England. Its population increased from 17,600 in 1779 to 50,000 in 1831,[10] owing largely to the insurge of impoverished rural workers, particularly the stockingers. Housing and living conditions deteriorated rapidly, owing to the impossibility of enlarging the town boundaries. Not until the passing of the Nottingham Enclosure Act in 1845 was any building possible beyond the town limits, and not before remorseless infilling, property speculation and the building of back-to-back houses in tight-packed courts had turned parts of the town into seething and noisome slums.[11]

On completion of his apprenticeship in his early twenties, George Green, father of the mathematician, went to work for a baker, and in 1791 he married Sarah Butler,[12] one of the four daughters of William Butler of Radcliffe-on-Trent. By now William Butler was a prosperous baker and property owner, and he may also have been George Green Senior's employer. Sarah's elder sister, Elizabeth, had married George Tomlin of Shelford six years earlier.[13]

William Butler's bakery was in Wheatsheaf Yard in the old quarter of narrow alleys running between Long Row and Backside (later known as Upper Parliament Street). Green probably worked for Butler until his marriage, then, aided by his father-in-law, he set up in business on his own in the east of the town, acquiring newly built property at the corner of Platt Street (now Lower Parliament Street). It was probably here, if not in her mother's house in Wheatsheaf Yard, that Sarah Green gave birth to her son George on 14 July 1793.[14] A daughter, Ann, was born in 1795; a second daughter died in infancy.

George Green Senior, after his marriage in 1791 at the age of thirty-three, appears to have prospered in business. On his death nearly forty years later, he was able to leave sufficient assets to permit his son to abandon the trade of milling and, as a gentleman of means, devote his remaining years to mathematical study. In 1792 Green purchased a second plot of land adjoining the one he had already acquired at the junction of Platt Street and Millstone Lane, on which stood the bakery and the family house.[15] On this plot a few years later he built sixteen back-to-back houses. These became known as Meynell Court; their outer frontages faced Meynell Row and Charles Street respectively. The construction of such courts, with a communal privy at one end and a low, tunnel-like passage at the other giving access to the street, became a feature of Nottingham housing and produced the appalling slum conditions condemned by the Royal Commission on the Health of Towns in England of 1845. The Byron ward, where the Green property was situated, was the worst of the seven town wards, with the highest infant mortality and life expectancy of only eighteen years.

Such conditions of overcrowding, poverty and deprivation provoked frequent riots in Nottingham. Bakers in particular were the targets of much violence during the Corn Riots at the turn of the century. The blockade on imports imposed during the Napoleonic Wars, poor harvests and the high price of grain made bread scarce and too expensive for the starving poor. Suspicion that farmers and bakers were hoarding grain so that prices would rise even higher caused riots, frequently recorded by Abigail Gawthern in her Diary (1757–1810), including one on 1 September 1800,[16] which provoked the following letter from George Green Senior:

> To the Right Worshipfull the Mayor and his Brethren haveing all my Windows Broke Last night and being much thretend to night with more mischief being done to me such as entering my House therefore Gentlemen I most Humbly Crave your protection such as the Berer of this Can explaine to you from your Humble servant George Green.[17]

It was not unknown for a baker to acquire his own mill, but it may have been with some idea of avoiding such troubles as a baker that Green decided to develop a second line to his business. In 1807 he purchased at auction, from John Musters Esq. of Colwick, an attractive plot of land 'situate at Sneinton . . . Commanding most extensive views of the River Trent, Trent Vale, Clifton Woods, Colwick, Belvoir Castle and a rich and highly cultivated country, diversified by numberless other picturesque objects'.[18] The sturdy and practically minded George Green may well have been less interested in the aesthetic attractions of the view than in the possibility of erecting on this commanding site a modern brick windmill in place of the old post mill which had stood there. The area surrounding the mill, still known as Green's Gardens, was subsequently let by their resourceful owner as garden plots to neighbours and the wealthier residents of Nottingham. Sneinton was then a village about a mile from the town, offering peaceful hours in the fresh countryside away from the congested and noisy atmosphere of early-nineteenth-century Nottingham. In 1812 the author of *Beauties of England* wrote: 'The village itself is rural, at present in some measure romantic: it has a number of pleasant villas and cottages, and has long been famous for a race of dairy people who make a very pleasant kind of soft cheese.'[19]

Greenwood's 1826 map of Nottingham shows four windmills in 'Snenton': thirteen on the Forest, two on the Carlton Road, one on Wilford Hill and others in nearby villages. Windmills were a familiar sight in the countryside and an essential feature of the economy. Those on Greenwood's map were mostly post mills, and all have disappeared. In building his 'brick wind Cornmill' (to quote from his will), Green showed his enterprise and business acumen. It was a five-storeyed tower mill, fifty feet high, twenty seven feet wide at the base, narrowing to fifteen feet at the summit. The base of the walls was twenty-three inches thick for half the height, narrowing to eighteen inches for the remainder.[20] The cap was fitted with a fantail, a device invented sixty years earlier which ensured that the cap turned of its own accord to keep the sails in the wind. There were four sails: two common (lattice frames covered with canvas) and two spring (shuttered, with a spring handle on the tip which opened and shut the louvres). There was a wide gallery from which the miller adjusted each sail in turn. Inside, the timberwork of oak, elm, deal, sweet chestnut and willow was carefully and soundly constructed. There were three pairs of millstones: two French burr stones (for grinding wheat into flour) and one pair of Derbyshire grits (for coarse grinding of barley and animal food).

The mill stood in a yard 'with Granaries, Stabling for eight horses, hay chamber, miller's house and tenement adjoining', to quote from the *Nottingham Mercury* of 18 October 1844 and the advertisement for a new mill tenant which appeared three years after the mathematician's death. It is a little unclear precisely what living accommodation was provided at the mill. The 'miller's house' is probably the Green family dwelling. The 'tenement adjoining' may have been the long, one storeyed, four roomed building built on to the side of the mill (below the gallery so that it did not take the wind from the sails), and traditionally known as the 'foreman's cottage'.

It might be appropriate here to describe the working life of a miller, since this was the mathematician's daily occupation from adolescence to his mid thirties. We know from Green's cousin William Tomlin that he found the miller's duties irksome,[21] but he had no alternative, since he was the only son, obliged to work in the family business. The same source tells us that whereas he had a frail constitution in youth, he later enjoyed many years of excellent health.

The miller's life was indeed a hardy one. He was out in all weathers, provided there was a wind to turn the sails. Work had to be done regularly to cope with the inflow of cereals and the outflow of flour and fodder. After a day or two of calm weather, it might be necessary to run the mill for long hours at a time and even at night. The operative floor of the mill was the gallery, or meal floor, and it was here that the miller had to spend most of his time. This floor gave access to the gallery outside, from which the miller could adjust the sails according to the wind. Too high a wind would turn the sails too fast; it was then necessary to open them partially and 'spill the wind'. In Green's Mill this meant partly furling the green canvas covering the two lattice sails and adjusting the lever at the tip of the two spring sails to open the shutters: the spring could be set, however, to allow a degree of automatic adjustment under wind pressure.

A second preoccupation was to ensure a regular supply of grain to the stones. Each floor had its function, and full use was made of gravity. Sacks of cereal were hoisted from the ground floor to the top bin floor. Grain was poured through a chute into the hopper on the lower bin floor, where the 'shoe' directed it into the 'eye' of the millstone on the stone floor below, whence the ground meal fell through a further chute into sacks on the meal gallery floor. A bell on this floor activated by the 'shoe', alerted the miller that the hopper on the floor above required replenishing, since it was essential that the millstones did not run dry: at best this damaged their cutting surfaces, at worst the heat generated through friction could produce a spark and the consequent danger of fire.

The inside of the mill was warm, dry and dusty, and a miller's work was thirsty work. The 'thirsty miller' was as well-known a figure as the 'dusty miller' ('The Dusty Miller' was the name of an inn on the Forest), and one may remember Chaucer's miller: 'very drunk and rather pale . . . straddled on his horse half-on half-off'.[22] It was also, at times, dangerous work: a miller could be injured by falls, or while adjusting sails in a high wind or fighting a mill fire.

Storms, which could damage sails, and lightning which could cause a fire, were always a hazard. An upstairs window which gave a view of the sails was later inserted in the side of Mill House. So even at night, as ever during the day, the miller's ear was attuned to the sound of wind, and he was ready at a moment's notice to avert possible catastrophe.

Many months of the year were occupied with milling the seasonal crops: wheat and barley for flour, oats and barley for fodder. Out of season there was a great deal of work to do on the mill: checking for damp and sealing cracks, painting, repairing ropes and pulleys and the varied and intricate parts of the mill machinery. Most important – and requiring continual attention – was the dressing of the inner faces of the millstones. Most mills had two or three pairs, so that only one set needed to be out of commission at a time. Mills were constructed by millwrights and the stones were dressed by itinerant stone dressers, but some of this work was often done by the miller himself.

Dressing the stones had to be done, on average, about once a month, and it was a skilled job requiring special tools, expertise and physical strength. The top stone, some four feet in diameter and weighing nearly a ton, had to be levered up and laid on its back. Such stones started off twelve inches thick, but after many years' use and dressing they might finish up only half that depth: a skilled eye could thus gauge the age of a stone. The patterns cut in the stone were traditional and precise, the basic principle being a scissor action as the runner stone turned on the bedstone. It was essential that the top stone was set straight and true, so that it ran evenly over the lower one.

Millwrighting and milling turned on the strength of men and horses and the skilled use of ropes, pulleys, wedges, levers, cogs, winches and windlasses. The lifting of the cap, weighing some eight or nine tons, to the top of the restored Green's Mill by a modern diesel-powered crane in 1981 seemed difficult enough, given the height and limit of the terrain on the top of Belvoir Hill. A similar feat in the early nineteenth century using only man- and horsepower would seem to have been impossible in those circumstances; indeed in the past it was the custom to construct the cap *in situ* on top of the tower.[23] A mill is one of the very earliest examples of technology, illustrating the practical application of both mathematical and physical principles. Although Green's scientific work was theoretical and did not involve practical demonstration, it may not be without significance that his earlier years of learning and reflection were spent in such an environment.

Having built his mill shortly after 1807, Green Senior must have employed a miller, since he continued to live in Nottingham. Some time before 1812 he and his family had moved to Goose Gate. He was then fifty-four and his occupation was given as 'gentleman', an indication of his increasing prosperity. He was served with notice regarding watching and warding,[24] a parliamentary measure by which responsible male householders served as guardians of peace during the frequent town riots (Robert Peel's police were not introduced until the 1840s). In addition to the Corn Riots, the Nottingham Luddites, in 1811 and 1812, fearful of unemployment in the face of mechanization, attacked the small factories the wealthier textile manufacturers were setting up and wrecked the stocking frames. George Green Senior, however, was exempt from duty because of his age, and his son, now aged nineteen, was not cited, as he was not a householder. The young George was made a burgess of the town at the age of twenty-one, qualifying by birth: his father had qualified through apprenticeship. As burgesses they had the right to a parliamentary vote, and voted Tory. They were also eligible for election to the Council, the oligarchic governing body of the town, but neither of them was sufficiently interested in local politics to become involved in the Corporation.[25]

In 1813 Sarah Green's father, William Butler, died at the age of seventy-seven. The following year Green Senior took over formal management of his father-in-law's bakery. This had been willed by Butler to his only son and eldest child, Robert, together with the family house and property in Radcliffe-on-Trent. Green leased the bakehouse in Wheatsheaf Yard from him for £40 a year for life.[26]

Finally, in about 1817, George Green Senior decided to move from the town to Sneinton. He built a family house next to the mill – now called Mill House, but known in the late nineteenth century as Belvoir Mount. Ann Green married her cousin William Tomlin in 1816, and moved into a large town house in Nottingham. Her brother George was now twenty-four and their father was rising sixty, so the time may have been propitious for a family move. The country air and open spaces of Sneinton would have been preferable to the close, overpopulated living conditions of Nottingham. Newcomers from the town had started to build in

'New Sneinton', below St Stephen's Church, as early as 1807,[27] and the Lunatic Asylum was established there in 1812. Belvoir Terrace and Notintone Place would be built in the course of the next twenty years. The Green family were some of the earlier residents; their growing prosperity was evident in their solidly built brick house with stone window facings, its two front reception rooms, five bedrooms, kitchen, larder and large garden with a wide view over Colwick, and on a fine day as far as Belvoir. The siting of the house was eminently practical: it was built to the south-east of the mill, where it would not interfere with the prevailing wind from the south-west, and a gate across the path up to the house enclosed the entire area of the mill-yard and house.[28]

Green Senior was not the only one to prosper. His daughter Ann was fortunate in her marriage to William Tomlin. He was the son of Sarah Green's elder sister, Elizabeth Butler, who had married George Tomlin of Shelford. George Green Senior had made his own way in the world, from young orphan apprentice to managing his own businesses. But despite his later promotion to 'gentleman' status, he appears to have remained a rough diamond, hard-headed, hard-working, and semi-literate. His nephew, by contrast, is accorded the status of gentleman from the first, despite the fact that the Greens and the Tomlins came from similar backgrounds. For several generations they had been neighbours, farming the Chesterfield acres and entering into property transactions. In 1809, for example, George Tomlin of Shelford and William entered a mortgage agreement with 'George Green of Nottingham, gent.' on land in Shelford, with subsequent arrangements up to 1824.[29]

George and Elizabeth Tomlin's eldest son, George, continued to work on the Shelford farm, as did his son Reuben and his family until the 1850s. The Tomlins' second son, William, and his younger brother, Edmund, however, established themselves in Nottingham. Edmund set up in business as a maltster, which would have been a profitable venture in view of the number of inns and alehouses in Nottingham and one, of course, with farming connections.

William, on the other hand, appears not to have been attracted to trade or craft. He was, it would seem, one who could handle money and make it work for him. From his early years he was involved in land and property transactions: in the 1840s he started to invest in railway stock, and when he died in 1890 he left a considerable fortune. On his marriage to his cousin Ann Green in 1816 he set up house in High Pavement, almost opposite the great town church of St Mary's, where they were married. By 1845 he was recognized as 'a very respected man, living on his property in this town'.[30] William and Ann had a number of children, four of whom survived into adulthood. Their eldest child – baptized George Green, presumably after Ann's father – died in infancy. Those who survived were Eliza Ann, born in 1821; William, born in 1827; Marion, born in 1829; and Alfred, born in 1837.

In about 1860 the Tomlin family moved from High Pavement to Belvoir Terrace in Sneinton, in the shadow of the mill, on land left to Ann on the death of her father. William Tomlin lived in No. 4 Belvoir Terrace with his wife, his two daughters and his younger son, and employed resident man and woman servants. His elder son, William, who practised as a surgeon in Sneinton and Nottingham, lived next door at No. 6. The Tomlins established themselves as an important family in the neighbourhood and their reputation lived after them. In 1911 it was recorded that 'The Misses Tomlin gave the clock in St. Stephen's Church tower and Miss Marion Tomlin did a great amount of good among the poor'.[31] When the later vicissitudes of the Green family in Sneinton are considered later in this book, the existence of these highly regarded cousins and neighbours should not be forgotten.

Old George Green died in 1829. By this time the miller and his family had also become a prosperous and respected family in Sneinton. In addition to his property in Nottingham and an interest in his father-in-law's business in Wheatsheaf Yard, Green owned his modern brick windmill, a substantial family house, and the considerable area of land surrounding them which he rented out as gardens. A George Green – most probably the old miller – is recorded as churchwarden of St Stephen's Church from 1820 to 1825, and a George Green was named in 1828 as one of the three trustees of a local charity.[32] The latter George Green was probably the mathematician, since the national inquiry into public charities from 1815 to 1829 aimed to regulate their administration, and younger trustees would have been preferred. In 1831 the younger George Green was appointed one of the overseers of the poor.[33] The rate on the family property, which was the second highest in Sneinton, indicates the standing and prosperity of the Green family at this time.

George Green Senior was buried in the family grave in St Stephen's churchyard, where lay his wife, Sarah, who had died four years earlier. His estate went to his two children, George and Ann. Ann inherited fourteen houses in Nottingham and the sum of £300, together with land on Belvoir Terrace in Sneinton – 'three roods of garden and orchard land with well, Garden house and all stock upon it'.[34] The well is still there, marking the site of the Tomlin property; it is now incorporated into the newly developed park surrounding the mill.

To his son, Green left 'my brick wind corn mill with the Granaries Miller's House Stable buildings and appurtenances near thereto . . . and all my right of interest in the Stock in Trade Carts Horses Bags and other Articles and Utensils used in such business along with my son'. In addition Green Senior left him 'the Mill Close containing two acres or thereabouts Also all those other two closes . . . containing together six acres or thereabouts with the new erected Messuage or Tenement and outbuildings lately erected thereon . . . now in the occupation of myself and William Smith (my miller)', together with seven houses in Charles Street in Nottingham. Two years before his father's death, the younger George Green had signed the formal declaration required of millers under Act of Parliament,[35] an indication of his increasing responsibility in running the family business. George Green, now aged thirty-six, had worked in his father's businesses as baker and miller for twenty-seven years. He would now have the means and the leisure to pursue his mathematical studies.

The facts so far recorded come largely from official sources, since there is all too little information about the personal life of the mathematician. Green himself reveals very little on the subject of his family, and that only indirectly in his will. George Green Senior, however, had made reference in his will to 'William Smith (my miller)'. William Smith had married Mary Smart in 1793 in the church of St Leodagarious in Basford: their daughter Jane was baptized in the same church in 1802.[36] An older daughter, Ann, was born in 1794. In 1802 William Smith is recorded in the Watch and Ward lists as living in Hill's Yard, Nottingham, only a couple of hundred yards away from Meynell Row where George Green Senior had had his bakery, so it is possible that young George had known Jane Smith since his youth.[37] When William Smith was appointed Green's mill manager he would have moved to

Sneinton, possibly to the foreman's cottage adjoining the mill. When Green built the mill, in 1807, his son was fourteen and about to start his apprenticeship as a miller. Thus young George Green and Jane Smith, nine years younger, would have got to know each other, or possibly resumed a childhood acquaintance.

Jane Smith was a lace-dresser[38] working in Radford, a village a mile or so to the west of Nottingham. In the 1790s it was a well-known beauty spot, but from 1800 onwards, like Sneinton a few years later, its population rapidly increased.[39] Early building was of good-quality houses with upstairs workshops for lace and hosiery workers, and Radford at this time was recognized as a 'superior working-class area'. Jane bore George Green their first child in 1824, the year after he joined the Bromley House Library, when he was thirty-one and she was twenty-two. The child was baptized in St Peter's Church, Old Radford, registered as illegitimate, and given her mother's name, Jane. Jane Smith exercised a similar discretion three years later when she gave birth to a second daughter, Mary Ann, by also having her baptized in Radford. In 1829 Jane was living in Harold Place off Windmill Lane in Sneinton, and her first son, George, was baptized there in St Stephen's Church. Jane's second son, John, born in 1831, was baptized in Radford, but her last three children – all daughters – were baptized in St Stephen's. Catherine and Elizabeth were born while Green was in Cambridge; Clara, the youngest of the seven children, was born after her father's return to Nottingham, and she was only thirteen months old when he died in 1841.[40]

Beyond these facts, there can only be surmise and conjecture as to why George Green conducted his family life in this fashion. Possible reasons will be discussed later, but it seems clear that Jane Smith realized early on that Green's family situation at home, or perhaps his later academic ambitions, were to preclude marriage. All the children were registered as illegitimate, and only Jane Smith's name appeared on the baptismal certificate, but after the birth of her first child, Jane gave all the subsequent six children the baptismal name of Green. There is evidence that after a few years the children dropped their surname of Smith and took the name Green, and in adulthood all were known as Green.[41] Jane herself acquired respectability after the mathematician's death, if not before, by becoming known as Mrs Jane Green. She is entered as such in her first appearance in the town directory – in 1844, when she was living in Notintone Place; in later years the number of the house was given as 3. Notintone Place was a recently built cul-de-sac of some thirty terraced houses, each with its front railing and small garden. Jane's neighbours included twenty or so of the 'gentry', who were listed separately in the town directory from the other inhabitants of New Sneinton. They included two clergymen; one was the curate of St Stephen's Church.[42]

Presumably the mathematician supported Jane Smith and her increasing family of children, and they were all well provided for in his will. But they lived, unrecorded and possibly unmentioned, in the background of his life. In the chapters that follow, they maintain a silent and unobtrusive presence until their father's death.

Chapter 2

George Green's Education

With the foregoing account of Green's family and working life in mind, anyone aware of the contemporary significance of Green's work is bound to ask how the miller developed into a mathematical physicist whose contributions to physics were acclaimed by leading figures in nineteenth-century science, and whose mathematics and scientific concepts are widely used in scientific research and in the technology of our own day. As Green's personal testimony is lacking – correspondence is scant and there are no memoirs, notes or diaries – recourse must be made to various landmarks indicated in the documented facts. It will be necessary, therefore, to investigate the surroundings in which Green found himself at different periods of his life, and to assess their probable effect on his development. Thus much of what follows is inevitably circumstantial and descriptive.

The most valuable piece of documentation is a letter written by Green's cousin and brother-in-law William Tomlin in 1845, four years after Green's death, in response to an enquiry from Cambridge, raised by William Thomson, later Lord Kelvin, on his discovery of the Essay of 1828. In the letter Tomlin says of his cousin:

> He lived with his parents to the termination of their lives and duly rendered assistance to his Father in the prosecution of his businesses which was firstly a Baker in Nottingham and afterwards a miller in Sneinton . . . [but] these assistances were irksome to the son who at a very early age and with in youth a frail constitution pursued with undeviating constancy, the same as in his more mature years an intense application to Mathematics or whatever other acquirements might become necessary thereto.[1]

To George Green Senior's credit, he did what he could to foster the abilities of his only son, and in March 1801 young George was registered as pupil No. 255 at Robert Goodacre's Academy in Upper Parliament Street, Nottingham.[2] Four terms later he left, and Tomlin gives as the reason: 'His schoolmasters soon perceiving his strong inclination for and profound knowledge in the mathematics and which far transcended their own, relinquished the direction of his studies . . .'. Thus four terms in school was the only formal education the future mathematician received, and at the age of nine he was put to work in his father's bakery. Any judgement of the adequacy of this period of instruction must be made against the background of the educational facilities available in the Nottingham of 1800.

At the turn of the century there were few schools of any standing in Nottingham – indeed few schools of any kind. The population then numbered about 28,800. A generation after Green, the situation in public education was well documented: about 60 per cent of children were receiving some sort of education in the town's twenty-seven 'Sunday Schools', run by the Church of England and various Nonconformist bodies. By then there had been a considerable expansion in education, aimed at providing religious instruction and the elements of literacy for the children of the poor and working classes. In 1802, however, a survey showed that only some 25 per cent of children were attending the half-dozen such denominational schools then in existence.[3] There is no indication of the number attending the 'town schools', also known as Sunday Schools, since Sunday was the only day of the week when children were not at work.[4] These had been set up by leading townspeople to promote literacy, and the moral welfare of the children of the poor.

Robert Raikes (1735–1811) had set up his first Sunday School in Gloucester in 1780, so that children could learn to read the Bible. His example was emulated in 1783 by a group of wealthy residents in Nottingham whose aim was to effect 'a reformation of manners in the lower ranks of mankind'[5] in an attempt to prevent crime. A town school was set up by public subscription in the Exchange Hall (now the Council House) in 1785, although any idea of reaching the poorest children was thwarted by the fact that the necessary conditions of attendance were clean hands and clean linen,[6] since at that time in the poor quarters of Nottingham water had to be paid for by the bucket.

Attendance at all these schools was sporadic, rarely lasting longer than a year or eighteen months – whether for economic or health reasons. Children of primary-school age worked for their living, especially in the stockingers' families. The stockinger worked at what was frequently a hired knitting frame set up in the family dwelling or in small workshops, fabricating the material from which stockings were made and sewn up. Hours were long, pay was low, and working conditions were often appalling; the stockinger's weekly wage was often less than ten shillings.[7] When employment was good, children worked in the family; when employment was poor, they might attend school. Many ailed and died. The later records of 1845 show, for example, that some 30 per cent to 45 per cent of children in the different areas of the town died before the age of four.[8]

In addition to the town and Sunday schools, there were two charity schools. The High Pavement Chapel Charity School, a Unitarian foundation, had been started in 1785. The Bluecoat School, established in 1706, was an offshoot of the Christ's Hospital Foundation in London, formed to succour foundlings and orphans. On leaving school the boys normally went into apprenticeship and the girls into domestic service.[9]

For Green Senior – owning his own bakery, his business prospering, and with an only son showing signs of unusual intelligence – the choice of school lay elsewhere. The existence of the Nottingham Grammar School of ancient foundation would suggest the likeliest possibility to a modern reader, since the grammar schools have been the backbone of middle-class education in England, a goal for those aspiring to middle-class status and a probable path to university for the bright boy. The Agnes Mellors Free School in Stoney Street, founded in 1513 and known in this century as Nottingham High School, was one of the oldest grammar schools in the country. The fortunes of these schools varied over the centuries. In 1800 the standard of the Nottingham Free School was low:[10] scarcely any scholars proceeded to the universities of Oxford and Cambridge, which were then the only universities in England, and at times it was barely able to recruit the complement of forty or so literate pupils.

The two headmasters during the period relevant to this study were the Reverend John Challand Forrest (1793–1806) and the Reverend John Toplis (1806–19). They were both graduates in mathematics from Queens' College Cambridge, with what would nowadays be first class honours, being ranked ninth and eleventh Wrangler[11] respectively. The pupils, all boys, were taught by the headmaster and the second master, or usher. The usher during Challand Forrest's and Toplis's time was the Reverend Robert Wood. Wood was a rather contentious personality, and neither headmaster had an easy relationship with him. On Toplis's return to Cambridge in 1819, Wood was made headmaster after twenty-seven years as usher and the Reverend Samuel Lund was appointed usher. Lund proved to have an equally unaccommodating character, and in 1833 both were pensioned off by the Corporation of Nottingham, who were the School's Trustees. Samuel Lund will make more than one reappearance later, as he was George Green's friend and executor.

Toplis appears to have done his best as headmaster. He complained to the Trustees that the school was 'grossly abused, not only in the management of its funds but in the Choice of Scholars' – some had not even been literate on admission. Toplis, however, had to face the charge from a parent and fellow clergyman of having pulled 'hair from the boy's head', but he confessed to being 'under irritation'.[12] Despite this, he had the reputation of being a kindly, fair-minded man, begging the Trustees for consideration for 'the Old Woman who cleans the Free School and who is lying very ill'.[13] In the light of this account of the Grammar School, it is unlikely that young George Green would have qualified as a free scholar, owing to his father's comfortable circumstances, and equally unlikely that his hard-headed father would have wanted to pay good money to send him as a fee payer to receive such a mediocre education.

We thus come to the last, the most popular, and at this time probably the best source of education for middle-class children: the private schools and academies. Such establishments had flourished in the preceding century; the better ones were run by schoolmasters or clergymen owning their own establishments and taking in boarders. They were often small, with some six or a dozen pupils. Children were sometimes sent away to school. Abigail Gawthern, one of the gentry of Georgian Nottingham, attended a young ladies' school in Clapham, London for three years; her cousin and future husband, Francis Gawthern, went to Birmingham at the age of sixteen to board with the Reverend Mr Brailsford, and 'to pay £25 a year'.[14] The schools in Nottingham were numerous, and varied greatly in size and attainments: Dearden listed seventy in his 1834 Directory. Subscribers to Green's Essay of 1828 included nine schoolmasters, of whom four were in Orders and six were listed as proprietors of their own schools. Some half-dozen of these schools appear to have prospered over a number of years. Byron is thought to have attended Mr Blanchard's Academy in St James Street. Mr Rogers, who later ran a prestigious establishment in Park Row, started in a rented room in a grocer's shop in Bridlesmith Gate, where Abigail Gawthern sent her son Frank to be his first pupil and paid him three guineas.[15] It is not clear whether there was any distinction between academy and private school, unless the former offered some elements of Greek and Latin, essential prerequisites for university entrance.

The two leading academies were the Nottingham Academy and the academy run by Robert Goodacre.[16] The first dated from 1777 and was situated in open ground on Back Lane, running parallel with Upper Parliament Street, in an imposing building recorded in a drawing by Richard Bonington.[17] It was taken over in 1804 by the Reverend John Blanchard, late of Park Row; and at one time it was attended by the Nottingham poet Henry Kirke White.[18]

Robert Goodacre's Academy in Upper Parliament Street probably owed its reputation to the extrovert personality and diverse activities of its proprietor-headmaster. With his son already showing strong interest in mathematics, Green Senior would probably have been drawn to Goodacre's Academy by the founder's strong interest in mathematics and the scientific speculation which was current at the time. Robert Goodacre is thus the first person known to have been involved in George Green's intellectual development. Although Green's actual schooling amounted to only four terms, it was not without significance, given the nature of Goodacre's personality and teaching. Robert Goodacre's activities as headmaster probably fostered, if not exceptionally developed, the child's 'intense application to mathematics'. Goodacre's enthusiasm for astronomy and natural science would also lead a young, enquiring mind into speculation about physical phenomena. His later career as a lecturer and popularizer of science reflects the cultural background of Green's adult life in Nottingham, and suggests the degree of support available to him for his adult work.

Robert Goodacre was born in 1777. After working as an assistant to a Mansfield schoolmaster, he set up his own school in Upper Parliament Street, Nottingham, when he was still only twenty. He married three times and had two sons, Robert and William, and a daughter.[19] There is little information on Goodacre's first school in Upper Parliament Street, which Green attended, but four years after Green left Robert Goodacre bought a plot of land owned by the Duke of Newcastle on Standard Hill, then an area of open country adjoining the Castle and outside the town boundary. Here he built a large three-storeyed building, with accommodation for some thirty boarders. A drawing by Thomas Barber, a Nottingham artist, shows the completed 'commodious building with an observatory at the top', as the school prospectus was later to describe it. Goodacre published his prospectus in 1808,[20] together with an *Essay on the Education of Youth.* Recourse to these will give some indication of the education provided by Goodacre, and it probably reflects reasonably accurately the principal features of the earlier establishment, despite the increased size and facilities of the new building. Therefore, with due caution, one may look at what Goodacre was offering in 1808 to glean some idea of what Green acquired in his four terms. It would appear that most pupils stayed only one or two years, and few stayed as long as three years or more.

Goodacre presented a curriculum in three sections, each of which was charged for separately at twenty to twenty-four guineas per annum, including board.[21] They covered Reading, Grammar, Penmanship and Arithmetic; Geography, Mathematics, Bookkeeping, English, Composition, Natural Philosophy, Astronomy, History and Biography; Latin and Greek. As an extra, pupils could be taught French 'by a very respectable native who lives a few doors from the Seminary'. The prospectus listed an extensive library, as well as charts and maps. Most important in Goodacre's estimation would have been his collection of 'Philosophical Instruments'. As these were reasonably small and transportable, some of them were probably in use earlier in Upper Parliament Street to illustrate the science teaching later described in the *Essay on the Education of Youth.* These included 'a universal compound telescope, an electrical machine, an air pump, a prism, a barometer, a thermometer, a quadrant, an orrery, a pair of twelve inch globes and instruments for surveying'.[22] Some pieces of apparatus, however, were denied Green: 'In a few months will be added telescopes and other instruments necessary for astronomical purposes, when the author's observatory is fitted for their reception.'

What is of particular concern in this account of Green's early years is the kind of mathematical and scientific instruction given by Goodacre. Whatever its deficiencies, he appears to be the only schoolmaster on record at this time offering such instruction. Goodacre's knowledge of mathematics seems to have been confined to arithmetic, bookkeeping and some elements of algebra and geometry. His expertise in the first two is evidenced by several publications advertised in the back page of the 1808 prospectus: *Arithmetic adapted to different classes of Learners*, already in its fourth edition; an *Abridgement to the above, adapted to the use of young ladies and others . . .*; *All the necessary Tables in Arithmetic and Mensuration*; and lastly, *A Treatise on Bookkeeping.* Some of these were highly successful publications; the ninth edition of the *Arithmetic* was published in London as late as 1839.[23]

For Goodacre's views on other branches of mathematics and on natural philosophy, one must turn to the *Essay on the Education of Youth*, also advertised in the prospectus. Goodacre's style of writing is fluent and generally moralistic in tone, as perhaps befits an enterprising pedagogue anxious to attract custom from affluent or ambitious parents and imbued with the contemporary idea of improving the mind and character of the young through education. He does not, he declares, 'study the interest of the boy but the embryo Man'. The study of science is useful here.

> Natural Philosophy is another branch of science eminently calculated to expand the mind. By the investigation of nature, a thousand charms are unfolded; whilst the recollection of having once been ignorant of these pleasing ideas, prompts the student to new heights, that the pleasure expected from them may not be lost through negligence.

After the rhetoric comes the science:

> To ascertain that the Air which we breathe . . . is in reality a fluid possessing the properties of weight and elasticity . . . that it is the medium of sound; – that wind is nothing more than air put in motion, and many other truths easily demonstrated by a pneumatical machine is a certain way to furnish matter for the contemplation and delight of any student not wholly absorbed in laziness and inattention.[24]

Robert Goodacre shared the contemporary interest in astronomy, as his enthusiasm for telescopes reveals:

> The science of Astronomy is admirably calculated to elevate the minds of youth, by a contemplation of the sublime and majestic . . . a planet moving round the Sun, as its common centre, and carrying with it 7 moons, yet at so great a distance that, with a motion of 22,000 miles an hour, it is still nearly 30 years in performing its revolution; – bodies whose period of return is only once in 450 years; – these and similar ideas involve the mind in wonder and reverential awe . . .

In the section 'On Geometry and the Mathematics':

> many useful and sublime truths can only be discovered by equations in Algebra, the use of logarithms, or the doctrine of the triangle and its reference to heights and distances. No particulars need be given respecting the mode of teaching these subjects; every person acquainted with them must know that they admit as little diversity in practice as any branch of science which comes under the notice of a Tutor.[25]

Little room for speculation here. If, in four terms, young George Green was given a good grounding in mathematical calculation and had his interest in physical phenomena aroused by Goodacre's rather naive enthusiasm, it may be just as well that he did not stay at the Academy for too long. The essence of Green's genius may be that it was allowed to develop free and untrammelled, both then and later. Had he, like Newton, gone to grammar school and on to Cambridge at the normal undergraduate age, we might not have had the great Essay of 1828 and the later investigations, with their striking originality and forward thinking. Goodacre, who has been allowed to speak in his own words, would not have presented his pupil with much of the science which is revealed in those works. But for the years when George Green was a young boy in Nottingham, it was fortunate that he came to Goodacre's school. Green's taste for mathematics and associated study, if undernourished, was not quenched; and Goodacre's personality, as well as probable association with his sons, could have been of considerable support to Green in later years.

Early in the 1820s, Robert Goodacre left the thriving Academy in the care of his elder son Robert, and spent five years in America as a lecturer in popular science, chiefly astronomy.[26] On his return in 1828 he continued to lecture in England, giving a series of lectures in Nottingham in that year, and finally dying on tour in Dundee in 1835. Robert Goodacre's younger son William took over the Academy, continuing his father's work as schoolmaster and popularizer of science. He himself gave lectures on chemistry and natural philosophy, and hosted those of visiting lecturers in his school for the benefit of his pupils and the general public. In this respect the Goodacres are a link with George Green's later intellectual development when he joined the Nottingham Subscription Library, a centre for such interest amongst the cultured public, where he found practical support for the publication of his Essay in 1828.

These events, however, lay in the future. Somehow an attempt must be made to bridge the gap between 1802, when the nine-year-old child left Goodacre's Academy to work in his father's bakery, and the thirty-year-old man who joined the Library in Bromley House. During that time, George Green made astounding strides in the study of mathematics. This in itself would be a matter of surprise, given the quality of mathematics employed in the Essay, and one explicable only in terms of genius and a capacity for self-education. The factor which creates the greatest surprise, however, is that Green employed mathematical analysis in studying the themes of the Essay, electricity and magnetism. Mathematical analysis was the term applied both to the concept and to the form of the calculus employed on the Continent, which derived from that evolved by Leibnitz. The latter's contemporary, Isaac Newton, had evolved his own form of the calculus, and this was the one used in England and, in particular, in Cambridge, Newton's own university and the home of English mathematics. If even Cambridge did not teach continental analysis and use the works of Laplace, Lacroix, Poisson and the rest, how did a provincial miller and untutored mathematician in Nottingham come across them and use them to such advantage?

When George Green Senior took his son away from Goodacre's Academy, he might have been disposed to pay for further instruction in mathematics, had a competent tutor been available. Had he not been so disposed, it is not inconceivable that young George himself might have paid out of his earnings at the mill. The most probable tutor would be a mathematician and a schoolmaster: such a one was the headmaster of the Grammar School. Challand Forrest died in 1806 when George Green was thirteen, and he may have taught the boy some of the traditional Cambridge mathematics. He was succeeded by John Toplis who, as headmaster, lived on the school premises at the top of Stoney Street, not ten minutes away from the Green bakery in Meynell Row. In about 1812 the family moved to Goose Gate, just round the corner from Stoney Street.[27] In 1812 Toplis was thirty-four; Green was eighteen. Given the latter's 'intense application' to mathematics and – as we shall see shortly – the former's active enthusiasm for the subject, it is possible that Toplis was Green's mentor. But to examine this possibility further, it will be necessary to go back a little in time and investigate Toplis's earlier career at Cambridge.

Chapter 3

Cambridge Interlude

This Cambridge review will cover the first quarter of the nineteenth century – a fruitful period for study. It will offer an opportunity for *un peu d'histoire* and an account of the mathematical background during these years, and permit further exploration of the possibility that John Toplis may have helped George Green in his mathematical studies. Finally, it will span the hidden twenty years of Green's life.

When Green joined Bromley House Library at the age of thirty, he was within five years of publishing his greatest mathematical work, yet there appears to be no record of how, as a provincial miller, he came to master the 'Mathematical Analysis' of the title, still less produce the profound and often original observations on contemporary physical problems. There are a number of mysteries in the life of George Green, but this is surely the most significant and the one to which it is most difficult to find an answer.

When Toplis resigned from the headmastership of the Free Grammar School in Nottingham in 1819, he resumed residence as Fellow at Queens' College Cambridge. In 1824 he was appointed to the College living of St Lawrence in South Walsham in Norfolk. He married, and died there in 1858 at the age of eighty-two. His widow outlived him by twenty-three years; on her death she left a sum of six hundred pounds, the dividend of which was to be distributed among the parish poor: known as the Toplis Charity, it still continues to be so.[1]

Toplis's sedate existence from middle to old age belies the fiery energy of his younger years at Cambridge. He was born in Arnold, then a village on the northern outskirts of Nottingham, and was admitted at Queens' College Cambridge as a sizar in 1797. His status as sizar indicates relatively humble circumstances, since it had originally involved the student paying his way by performing menial tasks such as waiting on fellow students at table and eating the leftovers, though the turn of the century saw the end of this servitude. Toplis graduated in 1802 and was ranked Eleventh Wrangler: he took his MA in 1804 and his BD in 1813, having been elected Fellow and Tutor of the College. His signature as Fellow first appears in the College 'Conclusion Book', in which decisions were recorded, in January 1811.

On his return to Cambridge in 1819 Toplis, now in his forties, was appointed Dean of Queens'. Toplis in Cambridge appears to have shown, as Fellow, more spirit than he did as headmaster in Nottingham. In 1821 he was involved in a dispute over seniority. He and two others had been elected Fellows on 12 January 1810, but for some reason Toplis, possibly absent from Cambridge on the crucial date, was not admitted to his Fellowship until the following week. Fellowships varied in standing and income and such a discrepancy could – and apparently did – prove significant. Toplis argued hotly against the acknowledged seniority of his colleagues, but to no avail.[2]

Toplis's connections with Cambridge were at a significant period in the history of the University in two respects, both of which are reflected in the person of the President of Queens', Dr Isaac Milner. In 1800 Cambridge was academically a backwater, but for some years it had been fiercely involved in a religious controversy in which the traditions of the Established Church were being challenged by the Evangelical Movement. The latter had no doughtier champion than Isaac Milner, and his considerable influence in Cambridge rested to a large extent on his advocacy of Evangelical principles.

From his earliest days in the University, however, Milner established a formidable reputation as a mathematician. The son of a Yorkshire weaver, he entered Queens' as a sizar, and graduated as Senior Wrangler in 1774. So impressed were his examiners at his mathematical ability that they added 'incomparabilis' after his name. Fourteen years later he was President of his College, a position he held until his death in 1820.

Milner was a big man in a number of ways. 'His vivacity of manner made him the observed of all observers', wrote Whewell of Trinity.[3] 'Dean Milner's appearance was exceedingly distinguished', wrote his respectful and admiring biographer, his great niece Mary Milner, 'He was above the usual height, admirably proportioned, and of a commanding presence. His features were regular and handsome, and his fine countenance was as remarkable for the benevolence, as for the high talent which it expressed'.[4] Leslie Stephen is slightly less respectful. 'In person Milner was tall, with a frame which indicated great bodily strength. . . . His voice predominated over all other voices, even as his lofty stature, vast girth and super incumbent wig, defied all competitors.'[5] His portrait confirms the assertion that at his death he weighed twenty stone. His burial in a lead coffin under the floor of the College Chapel must have been a memorable occasion.

Milner's career as a mathematician at Cambridge was perhaps rather less spectacular. He acted as moderator in 1780 and 1784 for the final Tripos Examination held in the Senate House, and he published four papers in the *Transactions* of the Royal Society, to which he was elected Fellow in 1776. Thereafter he gave courses in 'Chymistry' and maintained a lively interest in mechanics and scientific experiments. He was appointed to Newton's old Chair as Lucasian Professor of Mathematics in 1798, but he maintained no formal connection with the teaching of mathematics other than continuing as Examiner for the annual Smith's Prize.[6] In private, his interest in the subject was lively and lifelong. On his death he bequeathed his personal library of three thousand books, including over a hundred books on mathematics, to the College.[7] This would not be remarkable, were it not for the fact that some two-thirds of these were by contemporary French mathematicians: Laplace, Lacroix, Legendre, Poisson, Biot, to cite the best known. These volumes still exist in the Old Library at Queens', some of them leather bound, gold-titled, with Milner's bookplate on the inside cover; dates of publication range from 1794 to 1816, just four years before Milner's death.

Isaac Milner had travelled to France in 1784 and 1785 in the company of his lifelong friend William Wilberforce, later Lord Shaftesbury, and met some of the leading French mathematicians. This may have aroused his interest in their work and induced him to acquire, through Cambridge booksellers, copies of their books as they were published. 'His letters', wrote Mary Milner, who is long on her great uncle's theological interests but disappointingly short on his mathematical ones, 'allude to the interest with which he had looked over the mathematical works of some of the most celebrated analysts.'[8]

Milner's library also – naturally – included editions of Isaac Newton's works, especially of the *Principia* with commentaries, and the *Principles of Fluxions*. Newton, who died in 1727, was still considered the leading figure in English mathematics and science nearly a century later. Cambridge, which had welcomed him as student, Fellow and Professor, enshrined his memory and remained firmly wedded to Newtonian mathematics. Just as Handel's long shadow had discouraged musical growth for a century after his death, so did Newton's in the field of mathematics, with disastrous results for scientific progress.

> It is a subject of wonder and regret to many that this island, having astonished Europe by the most glorious display of talents in mathematics and the sciences dependent upon them, should suddenly suffer its ardour to cool and almost entirely to neglect those studies in which it infinitely excelled other nations.[9]

One might ask who wrote this lament, and be somewhat surprised to find that it was written by John Toplis, three years after graduation. Toplis was in all probability a protégé of Isaac Milner. The President of Queens', affable but despotic, was known for his personal bias in the appointment of College Fellows and Tutors. In the smaller resident communities of the period such as Queens', contact between the President and his chosen followers may well have been close, and Toplis, one feels, was one to make his presence felt. Milner could conceivably have fired and fuelled Toplis's enthusiasm for continental analysis and, indeed, given him the freedom of his library. There is, for example, an entry in Milner's handwriting in his personal inventory noting the loan of 'La Croix on the Differential Calculus 3 vols 4to taken out by Mr Joseph King'. If Toplis did not borrow Milner's books, he had the opportunity to order them from the Cambridge booksellers, who had long had an international trade.[10]

Toplis's judgement was ratified over a century later by Edmund Whittaker: 'The century which elapsed between the death of Newton and the scientific activity of Green was the darkest in the history of the University.'[11] Yet at that time Cambridge was the centre of mathematical study in England. Mathematics was the only serious subject of study, and academic achievement depended on Wrangler status. Even when the Classical Tripos was introduced in 1824, candidates, until 1850, still had to have taken the Mathematical

Tripos first.[12] Conservatism reigned, since homage to Newton was unremitting and there was no recognition of the changes in mathematics taking place elsewhere.

Cambridge appeared to be indifferent not only to important new work being published in France, but also to most of the activity taking place in England. Joseph Priestley in the late 1700s and John Dalton in the period under discussion made highly significant contributions to science. The Royal Institution was set up in London in 1799, with Sir Humphry Davy as its most notable figure, ably assisted by the young Michael Faraday.[13] This served the popular interest in elementary science by providing demonstration lectures to the educated middle class and some of the rising artisan class. Later in 1826 and 1828, King's College and University College, forerunners of London University, would be established and would offer a university education to all, including those debarred from Oxford and Cambridge because of their religious beliefs – or lack of them. The Royal Military Academy at Woolwich was another centre of mathematical study; it recruited several well-known mathematicians to its staff, including Olinthus Gregory and Peter Barlow, though the main thrust of its work was to train officers in branches of mathematics, mechanics and engineering applicable to warfare.

One might have expected the Royal Society to have promoted the advancement of science in similar fashion to that of the Académie des Sciences in Paris, which in fact had been founded in emulation of the London Society. Unfortunately, at this time the Royal Society, like Cambridge University, was at a low ebb. Founded early in the reign of Charles II, it had included Isaac Newton and Christopher Wren amongst its early members. It held its meetings in London, at which members' papers were read: these, and other papers communicated to the Society, were published in the annual *Transactions*, which also included an index and a list of 'Presents' – published papers sent by foreign – usually French – writers and the *Transactions* of similar societies abroad. The spirit of scientific enquiry which had led to the founding of the Society, however, though it was still present in a minority of Fellows, had been much weakened over the years. Dr Thomas Young, in his *Sketches of the Royal Society*, reported that not until 1820 were the majority of members practising men of science. Matters had not improved much in 1830, when Charles Babbage wrote his

Reflections on the Decline of Science in England, in which he heavily criticized the Society.

The number of Fellows was in excess of six hundred, much the same as today, but not all the Fellows – who included figures such as Lord Byron, for example – were knowledgeable in mathematics or science, since the Society functioned more as a gentlemen's club, with fees the main source of income. Election, therefore, presented few difficulties. It sufficed for the name of the person proposed to be entered on a certificate signed by three Fellows. This was displayed for ten weeks in the meeting room, and if no objection was raised, the nominee was automatically accepted. The Royal Society was thus weighed down by much dead wood. In addition, it suffered under the stultifying influence of Sir Joseph Banks, its President from 1778 until his death in 1820. Banks was distrustful of new ideas: 'He had a great dislike of the mathematical sciences and did all in his power to suppress them; a natural consequence of which is, that mathematical knowledge has greatly depreciated in England during the last forty years', wrote Olinthus Gregory to Heinrich Schumacher shortly after Banks's death.[14]

The Royal Society's lack of enterprise in London, and Newton's deadening influence in Cambridge, had together maintained the status quo, until – in the words of Whittaker, from the passage quoted above:

> A few years before Green published his first paper, a notable revival of learning swept the University: the Fluxional symbolism, which since the time of Newton had isolated Cambridge from continental schools, was abandoned in favour of the differential calculus, and the works of the great French analysts were introduced and eagerly read.[15]

The 'great French analysts' to whom Whittaker refers had established France at this time as the country to which 'all other nations turned for their scientific inspiration'.[16] One of the earliest pioneers of the French scientific renaissance was Charles Augustin Coulomb (1736–1806),[17] an engineer by profession, who published papers in electricity, magnetism and elasticity between 1770 and 1790: he is referred to frequently by Green. Joseph Louis Lagrange (1736–1813) published his *Mécanique analytique* in Paris in 1788, his *Théorie des fonctions* in 1797, and his *Traité des fonctions* the following year. Pierre-Simon Laplace

(1749–1827) started publishing the first books of his *Mécanique céleste* in 1799; they continued to appear until 1825. Adrien-Marie Legendre (1752–1833) published his *Traité de mécanique* in 1774, and Sylvestre François Lacroix (1765–1843) published his *Traité du calcul différentiel et du calcul intégral* in two volumes in Paris in 1797 and 1798. These seminal works, valuable sources for Green, had all been published by 1800.

After the turn of the century came further publications and younger men. Joseph Fourier (1768–1830) published his *Traité analytique de la chaleur* in 1822, and Jean-Baptiste Biot (1774–1862) brought out his *Traité de physique* in 1816. This gave a comprehensive account of contemporary physics and was of considerable value to Green. André-Marie Ampère (1775–1836) wrote on electrodynamics, publishing a *Mémoire sur la théorie mathématique des phénomènes électrodynamiques uniquement déduites de l'expérience* in Paris in 1827. Joseph Louis Gay-Lussac (1778–1850) worked on the thermal expansion of gases, and published in the journal *Annales de Chimie*, in 1802, an article 'Sur la Dilatation des gaz et des vapeurs'. Siméon-Denis Poisson (1781–1840) published his *Traité de mécanique* in two volumes in 1811, as well as papers on electricity and magnetism which were major sources for Green's Essay of 1828. François Arago (1786–1853) worked on electricity and magnetism, but especially, in collaboration with Ampère, on light. He was a protagonist for the undulatory theory of light, as opposed to the Newtonian corpuscular theory. Augustin Jean Fresnel (1789–1857) working almost exclusively in optics, finally demonstrated the correctness of the undulatory theory. In 1821 he wrote articles in the *Annales de Chimie* on the diffraction and the double refraction of light, and is quoted by Green in his memoirs on the subject. The collected works of Augustin-Louis Cauchy (1789–1857) comprise twenty-four volumes of papers, published posthumously in Paris from 1882 onwards. He is credited with the founding of the elastic theory and wrote classic works in hydrodynamics. His treatise on definite integrals (1814) provided the basis for the theory of complex functions; this was followed by a *Cours d'analyse* in 1821.

This catalogue of authors covers some fifty years from the 1770s onwards. It includes more than a dozen writers, known to international science, who produced innovative work in both mathematics and most branches of physics: electricity, magnetism, elasticity, hydrodynamics, heat and light. Some writers were both mathematicians and physicists, refining the mathematical techniques necessary for the accurate recording of experiments and for clarifying their theories of physical phenomena. Some, like Ampère, deduced theories from physical experiments. Fresnel, too, first conducted optical experiments and then produced a theory using mathematical analysis. Green, on the other hand, would proceed by mathematical analysis and then check his conclusions against the results of others' experiments.

Membership of the Académie des Sciences or the Société d'Arcueil, or working as colleagues at the Grandes Ecoles, brought French researchers into contact with each other in either collaboration or dispute, each a stimulant of active research. Some – like Lagrange, Laplace, Lacroix, Fourier and Biot – published books which enjoyed a wide circulation, and a considerable number of papers and articles. The most important papers were published as 'mémoires' by the Académie des Sciences;[18] others were printed as articles in reputable scientific journals such as the *Journal de l'Ecole Polytechnique*, the *Journal de Physique*, the *Annales de Chimie* or the *Bulletin de la Société Philomatique de Paris*. In the course of time, abstracts in translation appeared in the foreign journals: in Britain, for example, in the *Quarterly Review of Science and the Arts*, the *Quarterly Journal of Science and the Arts*, the *Edinburgh Review* or the *Annals of Philosophy*.

Other men of science – such as Coulomb, Ampère, Arago, Fresnel and Cauchy – published no major work in book form, but an immense amount of material in the shape of memoirs, articles and lecture notes. There was also a vast amount of correspondence, especially with fellow workers abroad, which included – incidentally – a few of the more enlightened British men of science: John Frederick Herschel corresponded with Legendre, Thomas Young with Arago. There were also exchanges with eminent philosophers in Germany and elsewhere: Lagrange corresponded with Euler in Berlin and Legendre with Jacobi in Königsberg. Jacobi's lifelong friend, Dirichlet, spent some time in Paris in the 1820s; Humboldt in Potsdam and Gauss in Göttingen had many contacts abroad: Gauss in particular left thousands of letters to posterity.

This correspondence was sometimes supported by personal acquaintance following a visit or a period of study abroad. Fresnel spent some time in Geneva and was a corresponding member of the Royal Society. Cauchy, after the July Revolution of 1830, exiled himself to Turin and Prague. Herschel, probably through his German connections,[19] was a corresponding member of the Academy at Göttingen. He and Babbage submitted a paper, printed in the 1825 *Transactions of the Royal Society*, on the 'curious experiments of M. Arago, described by M. Gay-Lussac during his visit to London in the spring of this present year. . . .'

Recent political events in France favoured this outburst of scientific activity. Although the 1789 Revolution largely destroyed the privileges of aristocracy and clergy, it did not destroy institutions valuable for educational training and the intellectual life of France, in particular long-established ones such as the Collège de France, founded in 1529 by Francis I; the Académie Française (1635) and the Académie des Sciences (1666), both founded by Louis XIV. The *ancien régime* had also introduced the first official attempts to promote technical training in setting up the Ecole des Ponts et Chaussées and the Ecole des Mines,[20] as well as the Ecole d'Artillerie and the Ecole de Génie Militaire. This development continued during the revolutionary period with the formation of the Ecole des Arts et Métiers in 1792, and the Ecole Polytechnique, the Ecoles de Médecine and the Ecole Normale Supérieure in 1794. Finally Napoleon founded the Ecole Militaire, which moved to St-Cyr in 1808; in the same year he established a comprehensive system of state education under the aegis of the Université de Paris.

Education and training in France was free, secular, and open to all on merit. The Grandes Ecoles thus took the best intellects, subsidized their research and publications, and provided entrants trained academically and technically to the major institutions and industries of the country. Most – if not all – of the French men of science under discussion were connected with at least one of these establishments. Some, like Ampère, Gay-Lussac, Cauchy and Poisson, held public posts in the Ecole Polytechnique, which enjoyed, then as now, the prestige of a great university,[21] or in one of the other Grandes Ecoles, in the Collège de France, or, like Arago, in the Paris Observatoire. Appointment to public office, therefore, provided a livelihood, facilities for experiment and research, and a platform for publications. Research was stimulated by the Académie des Sciences announcing a topic for a prizewinning essay. Fourier's *Traité analytique de chaleur* (1822), for example, was the eventual outcome of his winning the Académie's prize in 1810.

This desirable state of affairs in France was not lost on John Toplis, stranded in the scientific backwater of England. His 1805 article continues:

> The sciences are so abstruse that to excel in them a student must give up his whole time and that without any prospect of recompense; and should his talents enable him to compose a work of the highest merit, he must never expect by publishing it, to clear one half of the expense of printing.

The situation in England was indeed different from that in France. The English Revolution of 1689 – a century earlier than the French Revolution – was important for constitutional issues, but it left the fabric of English society intact. The colleges of Oxford and Cambridge were of clerical, royal or aristocratic foundation, some as early as the fourteenth century. Each college was a separate independent body, with its own traditions, endowments and autonomy. The University awarded the degrees, but the buildings, discipline, much of the teaching and most economic and social issues were the responsibility of the College. In Cambridge, provision by the University of requirements such as specialized libraries, and facilities for scientific research, were established only with difficulty in the second half of the nineteenth century, as colleges were reluctant to lose any independence or pool resources for the common good. An institution such as the Ecole Polytechnique in Paris, with all the necessary facilities for several hundred students, did not exist in England in the period under discussion.

Furthermore, the English colleges were institutions of privilege, largely accessible only to the affluent. For the rich, the lazy or the unambitious the universities were viewed more as a finishing school or an alternative to the Grand Tour. The value of a university education for the serious student was generally thought to be time for training the mind, cultivating the qualities of thought and character valuable in the liberal professions, or for high office in politics or public affairs. Subjects considered most useful to this end were the classics and mathematics, the preponderant subjects at Oxford and Cambridge. In the case of Cambridge mathematics, study was entirely theoretical, and the final results were submitted solely on paper; experiments or laboratory research formed no part of the course.[22] Finally, graduation at Oxford and Cambridge was restricted to Anglicans, so Dissenters, Non-conformists, Unitarians and Jews were virtually debarred from a university education until the foundation of the London colleges in the late 1820s.

On almost every count, therefore, Cambridge in particular and Britain in general lagged behind Paris and France, and failed to meet the challenge presented by current events in Europe. In France, a trained professional class of scientists, engineers and administrators, representing the best intellects in the country through competitive entry to the Grandes Ecoles, were assured of appointment to important offices of state, and in education and industry. Few such openings were assured for university graduates in England. A career outside the university – in politics, the law and other professions – depended to a considerable extent on family, fortune or influence; and for scientists, a career other than an academic one hardly existed during the early years of the century. In consequence, the main avenue to advancement for the able Cambridge Wrangler was in the University and in this way the system was perpetuated.

It might appear remarkable that England and France, geographically so near, remained so far apart in scientific and mathematical matters. In the rest of Europe – between France, Germany, Switzerland, Italy and Russia – there was frequent and fertile communication through university appointments, visits and correspondence. France and England, however, had been traditional enemies for centuries; the Revolution and the rise of Napoleon had represented a recent and very real threat to the security and social stability of England. The Napoleonic Wars and the economic blockade against France, which came to an end only in 1815, impeded – if it did not entirely prevent – normal intercourse between the two countries. This was yet another reason why England was cut off from intellectual contact, and thus from knowledge of the prolific and important discoveries of the French *philosophes*. What appears now to be a storm in a teacup, but caused a long and lasting furore between England

and the rest of Europe – and, in particular, Germany – was the controversy over who first formulated the calculus, Newton or Leibnitz. This was the cause of the 'War of the Philosophers', the consequences of which resounded for a hundred and fifty years, to the great disadvantage of the English.[23] The long-term result of this dissension for the mathematics of 1800 was that England remained locked into the Newtonian calculus, with its particular notation, while the continental mathematicians circulated their work using the calculus of Leibnitz, with his notation, which in the event proved more flexible. Continental mathematics on the Leibnitzian model eventually became known as mathematical analysis.[24]

It will now be realized how unusually progressive Isaac Milner was in his mathematical interests. He made no attempt to divert the course of Cambridge mathematics from the hallowed path of Newtonian tradition, however, and died in 1820, just as the mathematics used on the Continent began to be adopted in Cambridge. One or two significant events pointed the way. The first was the publication by Robert Woodhouse, Fellow and Tutor of Gonville and Caius College, of *The Principles of Analytical Calculation* (1803), the first book published in England to use the continental notation.[25] This was followed six years later by *The Principles of Trigonometry*. Both were used as university textbooks and helped, in the course of the first two decades of the 1800s, to bring about a revolution in Cambridge mathematics. Woodhouse is not known to have introduced mathematical analysis into his teaching, and his influence here seems to have been confined to his books. Charles Babbage had cut his teeth on them before coming up to Cambridge in 1810, and this provides a link with another attempt, similar to Toplis's, which will be discussed shortly.

John Toplis took his enthusiasm for French mathematics with him when he went to Nottingham in 1806. His article in the *Philosophical Magazine* a year earlier was no irresponsible diatribe. With missionary zeal he started on a translation of the first book of Laplace's *Mécanique céleste*, which had been published in Paris in 1799. Toplis published *A Treatise upon Analytical Mechanics . . . Translated and Elucidated with Explanatory Notes by the Rev. John Toplis, B.D.* in Nottingham in 1814, to which he added a Preface:

> It has been for some time a subject of complaint amongst mathematical readers, that, although analytical sciences have been investigated with the greatest ardour for a length of time by men of the most eminent talents upon the continent, yet scarcely any works exist in the English language in which the improvements made by them are noticed.
>
> As their notation and peculiar modes of proceedings are different from those used by English Mathematicians; I conceive that a translation of an elementary treatise upon analytical mechanics by one of the most distinguished of the continental analysts, with notes that shall enable the reader to understand it with greater facility, will render an acceptable service to those who are desirous of being in some degree acquainted with their merits.

Toplis proceeded to recommend further works 'upon the integral calculus', including those by Legendre and Lacroix, in particular the latter's treatise on the differential and integral calculus, as well as Lagrange's *Mécanique analytique*. 'With respect to the notes,' he continued, 'it may be proper to observe that they are intended to facilitate the reading of the text to those students whose information is not supposed to extend beyond the elementary principles of mechanics and of fluxions as taught in this island.'[26] Toplis's earlier complaint of want of patronage in England, such that the writer of a scientific work 'must never expect by publishing it, to clear one half of the expense of printing', was, in his case, prophetic, since two hundred and forty-seven copies of his book were found in his house after his death.[27]

Toplis's brave effort appears to have been largely abortive. Had he published in Cambridge, his book might have had wider influence. But in Nottingham, in the light of the circumstances as we know them, its effect must have been almost nil, except for its possible influence on George Green. This chapter began with the suggestion that John Toplis could have been George Green's tutor. It is obvious that Green could have bought a copy of Toplis's translation of Laplace in a Nottingham bookshop, and subsequently ordered the further texts recommended in the preface. In the light of the information on Toplis given above however, and the physical proximity noted in the previous chapter, it may perhaps be conceded that the suggestion is a valid one, and that John Toplis may have been the guiding hand that set George Green on his mathematical path to the Essay and beyond.

John Toplis was not the only one who had the idea of making analytical mathematics more accessible to English students. Two years after the publication of his translation of Laplace in Nottingham, the translation of an equally important work, Lacroix's *Traité du calcul différentiel et du calcul intégral*, was published in Cambridge. This was the result of a concerted effort by the Analytical Society, formed by a group of young undergraduates who shared an enthusiasm for French mathematics. It started as something of a joke, perpetrated by the young Charles Babbage.

Babbage was the son of a prosperous Devon banker; he received the usual education of a boy of his class and background. Largely self-taught in mathematics, he read several significant mathematical works during his schooldays, including Woodhouse's *Principles of Analytical Calculation*, published some five or six years earlier, and Lagrange's *Théorie des fonctions*.[28] Inspired by this introduction to continental mathematics, Babbage acquired a copy of Lacroix's book on the calculus in October 1810 on his way through London to Cambridge, where he enrolled at Trinity. He later moved to Peterhouse (then known as St Peter's), and took his BA in 1814, though he did not subsequently take the Mathematical Tripos. On arrival in Cambridge Babbage requested elucidation on various points in his French reading from his tutor, Hudson, who – good Newtonian that he was – advised him to concentrate on the fluxional mathematics required for the examinations.[29]

Babbage soon found himself associating with other lively minds also enthusiastic for the new mathematics, including John Frederick Herschel, son of Sir William Herschel, the Astronomer Royal; George Peacock; and Edward Bromhead, son of a Lincolnshire baronet. The Evangelical Revival, led by the renowned Reverend Charles Simeon and supported by the equally renowned Dr Isaac Milner, was then in full swing. One aspect of the controversy centred on whether the Bible should or should not be printed with explanatory notes, the proposal to do so being regarded by its opponents as profanity. One Sunday evening in his second year, Babbage picked up a controversial pamphlet of which he wrote a parody, substituting Lacroix for the Deity and his book on the calculus for the Bible.[30] He showed it to his friend Edward Bromhead, who, intrigued by the analogy, proposed the formation of a society to promote the new mathematics. A meeting was promptly held in Bromhead's lodgings, and the Analytical Society was formed with some dozen undergraduate members, including Babbage, Bromhead, John Herschel and George Peacock.

The Analytical Society published its first and only book of memoirs or articles – written, in fact, by Babbage and Herschel – the following year, 1813, when Bromhead graduated and left Cambridge. Bromhead realized, however, that the *Memoirs* 'were too profound to do us any good and not one mathematician in 1000 can understand them',[31] and that a full English translation of Lacroix's book on the calculus was needed. This was in fact provided by Babbage, Herschel and Peacock, and published in 1816; they followed it four years later with a book of worked examples in the new notation. The fact that Bromhead remained a guiding spirit during these years is indicated by his holding a meeting to draw up a constitution of the Society at his home in Thurlby Hall, Lincolnshire, in 1817.[32] Bromhead, then aged twenty-seven, was in the chair, and presumably host to the members.

After the publication of its third book, the Society was heard of no more. Of its active members, Babbage, Herschel and Bromhead all left Cambridge on graduation, as did the remaining half dozen or so, including Augustus De Morgan, Frederick Maule, and Alexander D'Arblay. Babbage and Herschel had been largely responsible for the three publications, with Peacock an active collaborator on the translation of Lacroix in 1816, and Bromhead an active supporter and critic throughout. A year later, Peacock's contribution was even more significant. At twenty-five he was appointed examiner for the Mathematical Tripos, and had the audacity to set his questions in the continental notation. This was a revolutionary move, as William Whewell was quick to note,[33] but against all expectation, Peacock was again asked to be examiner in 1819, and from this date the new notation took over. The last broadside from the old school came from the Reverend D.M. Peacock (no relation), who asserted: 'the Fluxional notation ought to be retained out of regard to the memory of the immortal Newton'.[34] Peacock and Whewell, both Trinity men, confirmed the victory on their appointment as Assistant Tutors by using the new notation in their teaching, thus distinguishing themselves from Milner and Woodhouse.

Peacock's action in setting Tripos papers in the continental notation in 1817 and 1819 suggests that in various ways French works had recently been filtering into Cambridge, and that undergraduates were becoming familiar with the new notation. This is borne out by the course reading prescribed for those years, as given to J.M.F. Wright, a student at Trinity from 1815 to 1819: the first year included Woodhouse, 'Plane Trigonometry'; the second year included Lacroix, 'Algebra' and Laplace, 'Système du Monde'; the third year Lagrange, 'Mécanique Céleste' [*sic*], Lacroix, 'Fluxions', 'Poisson', and the 'Journal Polytechnique'.[35] Woodhouse's books of the previous decade were obviously bearing fruit.

So far, this review of mathematical studies has been confined to Cambridge. Cambridge, as an establishment governed by tradition, maintained its hold on mathematical development until the early nineteenth century and the period we have just reviewed. Outside the University, a few individuals were more alert. We know of Herschel and Babbage, and we shall see that Bromhead kept abreast of developments; Augustus De Morgan, another original member of the Analytical Society, occupied an influential position on his appointment as the first Professor of Mathematics in the newly founded University College London in 1828.

Some well-known men – not trained in Cambridge – were at work elsewhere in the British Isles. William Rowan Hamilton, born in 1805, read Newton's *Principia* at the age of fifteen; two years later he discovered an error in Laplace's *Mécanique céleste*. When he went to Trinity College Dublin in 1823, he entered a department which had embraced continental mathematics since 1812, and where the French authors were used as textbooks. At twenty-two he was appointed Director of the Irish Observatory, and from 1837 to 1845 he was President of the Irish Academy. His compatriot and fellow student at Trinity College was Dionysius Lardner, who in 1829 initiated and edited the *Cabinet Cyclopaedia*, containing articles on electricity, magnetism and meteorology, to which Green referred in a letter to Bromhead.[36] James MacCullagh, four years younger than Hamilton, was appointed first Professor of Mathematics in 1836, then Professor of Natural Philosophy in 1843, at Trinity College Dublin. He was also Secretary of the Royal Irish Academy. In the 1830s he was working on the theory of light at the same time as George Green was publishing his papers on the subject in the Cambridge Philosophical Society's *Transactions*.

Sir David Brewster, born in 1781, and William Hamilton, born in 1788, were the best-known Scottish philosophers of this period, preceded by John Leslie and followed by James Forbes and W.J.M. Rankine. Brewster wrote prolifically on light, publishing papers in the *Transactions of the Royal Society*; he was President of the Royal Society of Edinburgh and founder-editor of the *Edinburgh Review*. He had numerous contacts with France and was made a foreign member of the Académie des Sciences. His compatriot, James Ivory, was an early convert to mathematical analysis, as his papers published in the *Transactions of the Royal Society* show. He used the continental notation in his first paper of 1809, 'On the Attraction of Homogeneous Ellipsoids', and there were further papers in 1812, 1814 and 1822. In his 1823 paper on 'Astronomical Attractions' he quotes Laplace, Gay-Lussac, Biot and Arago. Familiarity with the work of the French mathematicians earned him a place with Laplace, Lagrange and Legendre in 'applying infinitesimal calculus to physical investigations'[37] – as also, of course, did Green.

Against this background of enquiry and activity – from the Dublin mathematical revolution of 1812 and James Ivory's paper of the same year to his later papers of the 1820s – the prejudice against continental analysis and the rigidity of the Cambridge establishment, with the one exception of Woodhouse, are all the more striking. The lasting achievement of the Analytical Society was to prepare the way for the formation of the Cambridge Philosophical Society in 1819. Bromhead had no doubts as to who had fought the battle. 'The A.S. will always in history be called the parent of the C.P.S., as the meetings at Oxford are allowed to have been the origin of the Royal Society', he wrote to Babbage on 7 March, 1821.[38] Peacock, Whewell and Bromhead all became members, though Bromhead had returned to the family estate in Lincolnshire some seven years earlier.

The actual formation of the Cambridge Philosophical Society was due to the initiative of two men, Adam Sedgwick, Professor of Geology, and John Stevens Henslow, soon to be appointed to the Chair of Minerology. A public meeting was called at the University Library on 2 November 1819, and the first meeting was held on 13 December. By the end of the following year a hundred and seventy-one Fellows had been admitted, and £300 had been invested. Honorary members elected included John Frederick Herschel, Joseph Banks, David Brewster and Jean-Baptiste Biot, followed later by Faraday, Ampère, Barlow, Gregory, and others.

The first *Transactions* were published in 1821, and immediate steps were taken to establish exchanges with other scientific organizations in Britain, Europe and America. The entry in the Minutes for 12 March of that year includes the resolution that 'Copies of the Transactions be sent to the Institut de France, the Astronomical Society, the Irish Academy and the Royal Society of Edinburgh'. This policy ensured not only the Society's prestige but also a steady growth in the size of the Library, which ultimately came to be recognized as a valuable source of scientific information for the University.[39] The *Transactions* contained only papers – duly appraised by referees – by graduates and Fellows; these, together with gifts of members' publications, offered access to the most recent investigations in mathematical and scientific subjects. It is significant that Bromhead, in choosing from amongst the 'learned societies' the one to which he would send Green's first memoir, sent it to Cambridge, though as a Fellow of both Societies he could have sent it to either London or Edinburgh.

The Minutes of the first meetings of the Society in 1820 show that papers were read by Herschel, Babbage, Whewell and Peacock. The first two had left Cambridge some years earlier. On leaving Cambridge, Herschel went to help his father, the Astronomer Royal, before travelling out to continue his researches from the Cape of Good Hope. Babbage later became involved with his calculating machines, which took him outside the sphere of purely scientific research. Whewell and Peacock remained in Cambridge – Whewell for the rest of his life and Peacock until the mid 1830s, when he became Dean of Ely, though he continued to play an influential part in University affairs. They both remained staunch supporters of the Philosophical Society. Whewell served as Secretary and foreign correspondent; Peacock, as Treasurer, was involved in negotiations for new buildings, and in the 1840s each held office as President.

Many events of importance for the future had taken place in Cambridge in quiet and unobtrusive ways during the first quarter of the nineteenth century – not least in that when Green went to study there in 1833, the mathematical ground had been prepared and continental analysis was the accepted form of mathematics. One of the pioneers, John Toplis, had left for South Walsham nine years before Green's arrival. Green, however, would meet William Whewell and, in all probability, George Peacock. Bromhead he would already have met, in circumstances to be described later. But of this, Green had as yet no inkling. Sustained by his passion for mathematics, though lacking intellectual companionship and denied time for study, he continued to work long hours at the mill, helping his father on the daily round.

Chapter 4

Bromley House Library and the Essay of 1828

George Green published his first paper, 'An Essay on the Application of Mathematical Analysis to the Theories of Electricity and Magnetism', in 1828. The originality of the work stems from Green's own genius, but it may be assumed that his admission to the Nottingham Subscription Library in his thirtieth year was of considerable significance, and gave him support in various ways. Green was a member of the Library for ten years, during which time he produced not only the Essay but three further investigations.

The Nottingham Subscription Library still exists. Housed in Bromley House on Angel Row, facing the Market Place, it was the centre of cultural and scientific interest in Nottingham by 1823. It was one of dozens of such libraries in England, some of which were much older than the Nottingham Library. Bookshops not infrequently housed a lending library, and there were several such libraries in Nottingham,[1] but the better educated amongst the town's population felt the need for a more exclusive organization which would offer the facilities of a club and would provide books of their own choice in more specialized categories, reflecting the character of the traditional gentleman's library. The Liverpool Lyceum, founded in 1757, was possibly the first such library to be established, and the Nottingham Subscription Library sent for a copy of its catalogue as a guide for their first purchases.[2]

The first step, however, was to solicit the patronage of the Duke of Newcastle, Lord Lieutenant of the County and owner of Clumber Park and Nottingham Castle. Letters were then sent to local Members of Parliament and 'respectable inhabitants of the town'.[3] One hundred and fifty people replied. Subscribers purchased a share for five guineas and paid a two guinea annual subscription, with an additional fee for the use of the Newsroom. 'Trustees' were appointed as a committee under the chairmanship of Dr John Storer, FRS, a leading physician and a man much respected in the town.

After first meeting in Wright's Bank in Carlton Street, the Nottingham Subscription Library bought Bromley House in Angel Row in 1822. Bromley House, built in 1752, was 'an elegant brick building with iron palisades before it'.[4] The front door led into a large Entrance Hall, with the Newsroom on the left and a wide area on the right, separated by four columns and containing a large fireplace. A door in line with the front door opens on to a garden at the rear of the house, which is still visible through the open doors on fine summer days. A wide staircase leads up to the first-floor rooms where the Library is now housed, and where a later committee installed a delicate iron spiral staircase, which gives access to the upper shelves and to two further storeys, the attic storey providing views of Long Row and the Market Place from the dormer windows.

During this century pressure built up to lease the ground floor for shops (the basement had always been leased, formerly as wine vaults and latterly as basket stores). In 1916 the then President, Mr John Russell, in his history of the Library published to mark its centenary, recorded the rejection of the proposal:

> for fear of wounding the feelings of the older members of the library and from a desire not to mutilate a fine old building and destroy one of the last relics of a time when [in the words of Blackner, writing in 1815] 'the gentry of the county resorted to their county town for the winter festivities instead of going to London'.[5]

In 1926, however, the drastic – albeit economically necessary – decision was taken. It was a time of great national depression, coinciding with the removal of the market stalls from the Market Place and hence the need for new shops. As a result the Georgian façade is irretrievably mutilated, but the library is still functioning, having acquired a permanent income through leasing the ground floor. The Newsroom became one shop, the spacious area behind the columns another (they are still visible embedded in the dividing wall), and all that remains at street level is the charming eighteenth-century doorway, easily overlooked between two shop fronts.

Bromley House soon became a centre for many of the cultural activities of the town. The Minutes record a large and varied number of societies and individuals who sought to hire rooms in congenial surroundings. A Debating Society and a Literary and Debating Society were soon formed; the Law Society took rooms; the Geographical Society and later the Nottingham School of Medicine held meetings there; billiards clubs and chess clubs hired accommodation; there were meetings of the Amateur Musical Society, the Ladies' Committees of the Bible Society and the Female Artizans [*sic*] Society; there were dancing classes, art exhibitions and Florist Society displays, which were organized by the Bell Inn, a few doors down. An attempt at a further activity came from the enterprising William Tomlin. In 1827 he proposed at the Annual General Meeting that 'The Hall be appropriated to the Savings Bank by a temporary Skreen for three hours every Monday morning, and once a month on a Saturday', but 'the Proposition was put and negatived'.

The dramatic increase in population and the consequent deterioration of social conditions during this period were described in Chapter 1; Chapter 2 noted the efforts of philanthropical members of society, largely middle-class, to bring some benefits of education, or at least literacy, to the poor and deserving. A parallel development was a widening of curiosity in scientific matters,[6] and a growing public for local and itinerant lecturers such as Robert Goodacre. One of the earliest was the Quaker John Jackson, who delivered a well-received chemistry course in 1810, though John Nicholson had given similar lectures as early as 1806. Such lectures depended heavily on spectacular demonstrations. Jackson made use of some two dozen pieces of apparatus, including:

> Air Pumps . . . elegant working models of the Steam Engine . . . Modern electrical Apparatus . . . a superb cabinet of Optical apparatus . . . Telescopes . . . Magic Lantern . . . a very powerful and illustrative Magnetic Apparatus . . . a very accurate and elegant Orrery and Lucernal Transparencies for Astronomical Illustrations.[7]

After its establishment in 1816, Bromley House swiftly became a focus for this scientific curiosity and philosophical speculation, and commissioned speakers to give papers or courses of lectures. Their own members gave talks – for example, Mr Henry Oldknow, surgeon to the Infirmary, gave one on vision – and heated discussions took place in the Debating Society, all of which received due notice in the local press. The members took their scientific observations seriously. A device which still exists was installed in the rear upper room overlooking the garden: the sun, shining through a hole in the wooden shutter of the window, fell on a brass strip on the floor at an angle to give the precise time of noon in Nottingham. A wind machine was installed in the Reading Room, attached to a vane in the roof, allowing members to assess the force and direction of the wind. A pair of globes, maps, and an atlas were normal additions to a gentleman's library. At this time they would also serve as references in discussions on the controversial aspects of astronomy, warmly argued in the Debating Society by the Reverend Robert White Almond, who was also a member of the Astronomical Society,[8] or on reports of scientific naval expeditions, published in the *Transactions* of the Royal Society, which examined navigational problems such as the variations in the behaviour of the magnetic needle in northern latitudes.

Activities such as this indicate that on joining the Library in 1823, Green had contact with lively-minded people with a taste for scientific discussion. He would at least have had the opportunity to meet a wider circle of people than his family, his business associates and the occasional friend, such as Samuel Lund or Robert Goodacre. He would probably have found sympathetic support, for example, in the chairman, the Reverend Robert White Almond, who, like Challand Forrest and Toplis, was a Queens' man. Certainly Green was to find generous sponsorship from members for the publication of the Essay five years after his admission. It was probably here, too, that he found the 'kind and respected friends' mentioned by William Tomlin, who 'were anxious that he should adopt an University education several years before that circumstance actually took place, and [who] also assisted him in making known, and in the dedication of his first publication'. This psychological and practical support may have been the main advantage which membership of Bromley House conferred on Green. It widened his social experience and ended his intellectual isolation. Independent and reserved as he surely was, his passage into this new society was probably eased by the presence of his more worldly cousins, William and Edmund Tomlin; and Robert Goodacre Junior became a member the following year.[9] Yet it would be rash to assume that Green became a well-known and popular figure in this company. His own testimony, as well as another's, suggests that he was of a reserved disposition and possibly conscious of his own status as a working miller amongst these educated and professional people. Moreover, he would not have had as much time as these gentlemen of a more leisured class to browse in the Library and spend evenings in discussion. Such hours as he had free were more probably spent in consulting the books and journals relevant to his studies.

In spending £600 on books in the first months of the Library's existence, the committee naturally purchased those that reflected the current taste. It is of some interest to see the range of subjects listed under a traditional classification system, still in use today.[10]

Class A	Theology	
Class B	Philosophy	(a) Moral, Metaphysical and Logical
		(b) Natural and Mathematical
Class C	History	(a) History, Chronology and Biography
		(b) Geography, Topography, Voyages and Travels
		(c) Antiquities, Heraldry and Numismatics
Class D	General Literature	(a) Philology and Criticism
		(b) Poetry and Drama
		(c) Prose Works of Fiction
		(d) Polygraphy and Miscellanies
Class E	The Fine Arts	
Class F	Law, Politics, Naval and Military Tactics	
Class G	Periodical Publications	

The Minutes of the early years of the Library record spasmodically, and with some vagueness, the titles of books ordered. From 1816 onwards various catalogues were completed, but these were not strictly cumulative, and books were not uniformly accessioned with date of acquisition. Journals, as distinct from books, are recorded only occasionally in the Minutes, and one has to wait until the 1881 catalogue for a comprehensive retrospective list. Thus it is not always possible to establish with any accuracy which books were available to Green between May 1823, when he joined the Library, and December 1827, when the Essay was in press. Those of interest to him would be those modestly subsectioned in Class B (b), Natural and Mathematical Philosophy. In the 1820s these were few in number, but included the following (the date of acquisition is given when available):

The 1816 catalogue lists:

Hutton's Course of Mathematics, 3 vols; published in London, 1798–1801, by Charles Hutton, LLD, Professor of Mathematics at the Royal Military Academy, Woolwich;
Laplace's Astronomy, translated by Pond, 2 vols; published in 1809, by the Astronomer Royal;
First Book of the Céleste Mécanique [*sic*], translated by J. Toplis as a *Treatise upon Celestial Mechanics*; published in Nottingham in 1814. This was probably presented to the Library by the author during his brief three months' membership in 1816;
The Philosophical Transactions of the Royal Society: these included the abridged volumes from 1665 to 1800, and the annual publications thereafter.

Books recorded in the Minutes include:
A short Account of experiments and instruments depending on the relations of Air to Heat and Moisture, by J.S. Leslie, FRS; published in Edinburgh in 1813 and acquired in 1821;
Singer's Practical Electricity and Galvanism: this could be *Elements of Electricity and Electro-Magnetism*; published in London in 1814;
Leslie's *Philosophy of Arithmetic*; published in Edinburgh in 1817;
Hutton's *Mathematical and Philosophical Dictionary*; published in 1795;
Elementary Illustrations of the Celestial Mechanics of Laplace; published in London in 1821, by Thomas Young;
Treatise of Mechanics, by Olinthus Gregory, Hutton's successor at Woolwich; published in 1806;
Elementary Treatise on Mechanics, by William Whewell; published in Cambridge in 1819, but acquired only in 1828.

As Green joined the Library in 1823, with less than five years to go before completing the Essay, it seems unlikely that the more elementary books of this type would have been of much use to him at this stage of his development. Toplis's translation of Books I and II of Laplace has been mentioned already, though it is to Book III that Green consistently refers in the Essay. Hutton's and Gregory's books were written largely as textbooks for military cadets; Whewell's book came into the Library too late to be of value. An exception can be made in the case of the *Philosophical Transactions of the Royal Society*. Three papers are specifically referred to in the Essay: Cavendish's paper of 1771, Barlow's of 1825, and Thomas Young's of 1800. It is possible that other papers served Green as secondary sources, or as stimulus and background information, since one may assume that on joining the Library he was anxious to know to what extent mathematical analysis was used in England, and to learn as much as he could of experiments and research into physical phenomena. Thus papers by James Ivory, a Scot, who early in the century shared the interest in French analysis of Milner, Toplis and Woodhouse; and those by Davy, Woodhouse, Brewster, Faraday, Barlow, Babbage, Herschel and others, would extend the horizons of the self-educating miller.

Thus with the exception of the *Transactions of the Royal Society*, none of the books used directly by Green was to be found in the Library. Book III of Laplace is not there, nor are texts of Poisson, Biot or Coulomb, all of which are most frequently cited in the Essay; nor those of Arago, Fourier or Cauchy, to whom more general reference is made. H.G. Green analysed with some exactitude Green's sources, both those acknowledged and those implied.[11] He found them reduced to: Biot's *Traité de physique*, published in Paris in 1816; Lacroix's *Traité du calcul différentiel et du calcul intégral*, published in three volumes in 1810, 1814 and 1819; Poisson's *Mémoire de l'Institut* (1811; actually 'two memoirs of singular elegance', as Green calls them) and his *Mémoires de l'Académie des Sciences* (1821 and 1822), though Green also refers to Poisson's third memoir of 1823, in which he 'deduced formulae applicable to magnetism in a state of motion'.

So where did Green find these sources? The books by Laplace, Lacroix, Lagrange and Legendre were all recommended by Toplis in the preface to his translation of Laplace, and these, together with the Biot, could have been ordered through the Nottingham booksellers.[12] The references to Arago, Fourier, Coulomb and Cauchy are possibly from secondary sources: principally from Biot's *Traité de physique* but also from references in papers in the *Philosophical Transactions*. As for Poisson's memoirs on electricity and magnetism,[13] Green could have known of their publication through two sources: the *Philosophical Transactions* and the scientific journals. As we have seen, the annual publication of the former included a list of 'Presents' – papers and *Transactions* received from individuals and organizations in Britain and abroad. The regular receipt of the *Mémoires de l'Institut de France* did not, however, reveal their contents, and non-members were allowed access to the Society's Library only if they were accompanied in person by a Fellow, or on presentation of a letter from the President or a member of the Council.[14]

The second source of information about new publications was the notices and abstracts in the scientific journals. Poisson's two memoirs on electricity, for example, were noted in Thomson's *Annals of Philosophy* in 1813 and 1814, and these are listed in the 1884 Bromley House catalogues.[15] In 1825 the Library took out a subscription to *Brewster's Edinburgh Journal of Science*, founded the previous year by the eminent Scottish philosopher Sir David Brewster. It also subscribed to the *Quarterly Journal of Science and the Arts*, originally a publication of the Royal Institution but published under this title after 1816. Both these journals carried short abstracts of the later Poisson memoirs on magnetism: *Brewster's* limited to a couple of pages for each, the *Quarterly Journal* devoting ten and seventeen pages respectively.[16] It is perhaps just conceivable that the longer abstracts of these memoirs would have sufficed to launch Green on a fruitful line of enquiry. As for Poisson's third memoir on magnetism in motion (founded, as Green states in his preface to the Essay, on 'M. Arago's discovery relative to the magnetic effects developed in copper, wood, glass, etc, by rotation'), his insight into this topic may well have been stimulated by reading the paper submitted to the Royal Society in 1825 by two Fellows, Charles Babbage and John Frederick Herschel: 'An Account of the repetition of M. Arago's experiments on the magnetism manifested by various substances during the act of rotation'. What is not known is whether Green had access to the full text of the memoirs. His admiring reference to Poisson's two earlier memoirs 'of singular elegance' which 'to be fully appretiated [*sic*] must be read', suggests that he had. In the last resort, however, it may perhaps be assumed that Green was already far advanced along his own original path. To one of his genius, a single surmise might suffice to launch an idea: a detailed elaboration might be unnecessary.

Green was conscious of the paucity of his references, though this factor inevitably highlights the originality of his work. 'After I had composed the following Essay,' he states at the beginning of the preface:

> I naturally felt anxious to become acquainted with what had been effected by former writers on the same subject, and had it been practicable, I should have been glad to have given, in this place, an historical sketch of its progress; my limited sources of information, however, will by no means permit me to do so; but probably I may here be allowed to make one or two observations on the few works which have fallen my way . . .'[17]

It has been pointed out that there is a certain ambiguity in this passage[18] – the first sentence implies that the content owes nothing to any other author, yet Green made explicit reference to several authors of the French school of analytical physics. Possibly he felt that an 'Essay' should have had a historical preamble which he is unable to provide, for the reasons given. He was not, therefore, claiming full originality. It is unlikely that Green would be making extravagant claims for the Essay, in view of the diffident note struck in the final paragraph of his preface:

> Should the present Essay tend in any way to facilitate the application of analysis to one of the most interesting of the physical sciences, the author will deem himself amply repaid for any labour he may have bestowed on it; and it is hoped the difficulty of the subject will incline mathematicians to read this work with indulgence, more particularly when they are informed that it was written by a young man, who has been obliged to obtain the little knowledge that he possesses, at such intervals and by such means, as other indispensable avocations which offer but few opportunities of mental improvement, afforded.[19]

Green then proceeds to make 'the one or two observations' on the few works which have fallen his way. The first is on the paper by the 'celebrated Cavendish', printed in the *Philosophical Transactions* of 1771: 'An Attempt to explain some of the principal phenomena of electricity by means of an elastic Fluid'. Referring to one of Cavendish's propositions, he points out that 'a trifling alteration will suffice to render the whole perfectly legitimate'. He selects the twentieth proposition, the object of which is 'to show, that when two similar conducting bodies communicate by means of a long slender canal, and are charged with electricity, the respective quantities of redundant fluid contained in them, will be proportional to the "n–1" power of their corresponding diameters: supposing the electric repulsion to vary inversely as the "n" power of the distance'. He shows that Cavendish's proof agrees with actual experiments – provided, as demonstrated since by Coulomb, that 'n' is equal to two.[20]

Green notes that since the publication of Cavendish's paper, little had been done on the mathematical theory of electricity until the appearance of Poisson's two memoirs on electricity in 'about 1812'; these dealt with 'the distribution of electricity on the surfaces of conducting spheres, previously electrified, and put in presence of each other'. According to Green, Poisson's memoirs:

> are in fact founded upon the consideration of what have, in this Essay, been termed potential functions, and by means of an equation in variable differences, which may immediately be obtained from the one given in our tenth article, serving to express the relation between the two potential functions arising from any spherical surface, the author deduces the values of these functions belonging to each of the two spheres under consideration, and thence the general expression of the their actions on any exterior point.[21]

Since both 'the electric and magnetic fluids' are subject to 'one common law of action, and their theory, considered in a mathematical point of view, consists merely in developing the consequences which flow from this law, . . . it is evident the mathematical theory of the latter, must be very intimately connected with the former'. Considerable difficulties faced the analyst, however, when considering 'bodies as formed of an immense number of insulated particles, all acting upon each other mutually . . . and . . . until within the last four or five years, no successful attempt to overcome them had been published'. Here Green turns to Poisson and his three memoirs on magnetism. The first two contain 'the general equations on which the magnetic state of a body depends'; in the third, Poisson has deduced formulae applicable to magnetism in a state of motion, based on Arago's discovery of the magnetization of various materials in rotation. But as Green is quick to see, 'although many of the artifices . . . are remarkable for their elegance, it is easy to see they are adapted only to particular objects, and that some general method, capable of being employed in every case, is still wanting'.

Green sets himself to provide it. To do this, he has recourse to 'the use of a partial differential equation of the second order' from Book III of Laplace's *Mécanique céleste*. Combining this with his own previous work on these equations, he was able to apply the result 'with great advantage to the electrical theory' and 'demonstrate the general formulae contained in the preliminary part of the Essay'.[22] These have come to be known as Green's functions[23] and Green's Theorem, and are the kernel of the work. Parts two and three of the Essay 'ought to be regarded principally as furnishing particular examples of the use of these formulae' . . . first 'to the facts on which the electrical theory rests', and secondly to 'the hypothesis on which the received theory of magnetism is founded'. These:

> are by no means as certain as the facts on which the electrical theory rests; it is not the less necessary to have the means of submitting them to calculation, for the only way that appears open to us . . . is to form the most probable hypotheses we can, to deduce rigorously the consequences which flow from them, and to examine whether such consequences agree numerically with experiments.

This indeed will prove to be Green's blueprint for future work: to produce mathematical theories, and then to test them against the results of others' practical experiments.

'The applications of analysis to the physical Sciences', Green continues, 'have the double advantage of manifesting the extraordinary powers of this wonderful instrument of thought, and at the same time of serving to increase them.' Thus:

> M. Fourier, by his investigations relative to heat, has not only discovered the general equations on which its motion depends, but has likewise been led to new analytical formulae, by whose aid M.M. Cauchy and Poisson have been enabled to give the complete theory of the motion of the waves in an indefinitely extended fluid.[24]

Finally, if:

> astronomy, from the state of perfection to which it has attained, leaves little room for farther applications of [the analysts'] art, . . . probably the theory that supposes light to depend on the undulations of a luminiferous fluid, and to which the celebrated Dr. Young has given such plausibility, may furnish a useful subject of research, by affording new opportunities of applying the general theory of the motion of fluids.

In later years, Green himself would apply mathematical analysis to some of those phenomena.

The Essay remains, however, Green's *magnum opus*. As a result of the circumstances of its publication, it was a neglected work. French contemporaries would have welcomed it, but in England contemporary interest in the subjects of electricity and magnetism appeared to be lacking, or of an incidental nature. Babbage and Herschel published some papers in the *Philosophical Transactions*, but Herschel's main interest was astronomy, and Babbage was to devote his life's energy to the construction of his 'calculating engines'. It would be seventeen years before the value of the Essay was recognized, and by that time Green was dead. It would be twenty years before it achieved wider publication – and then abroad – and a generation before Green's work would actively contribute to the development of science in England.

Examination of Green's sources for the Essay suggests that an early study of Toplis's preface to his translation of Laplace may have been sufficient to put him on the track of Laplace, Lacroix and the French analytical movement generally, and Toplis's book was obtainable in Nottingham. Green obviously became proficient in reading the French texts.[25] Comprehension was facilitated by the common language of mathematics and also by the formal, Latinized vocabulary of the text, containing words of common root. Nor would Green be unfamiliar with the basics of French syntax, if he had perhaps followed the French course offered at Goodacre's Academy. There was in any case quite a strong French presence in Nottingham. Trade between England and France in the lace and textile industries led to an exchange of commercial and cultural interests;[26] and in January 1835 the Bromley House Minutes record 'A Petition having been presented by the Foreign Subscribers for a French paper . . . scarcity of funds oblige the Committee to return a denial'.

Booksellers played their part; they were able to procure French texts, and advertised them in the press. On 27 May 1814, for example, the *Nottingham Review* carried a notice of 'A new French and English Dictionary completely adapted for the English Visiting France and the French Visiting England. Just published, price Twelve Shillings Bound'; also 'Wanostrocht's Grammar of the French Language 4s 6d bound'. Finally, this was an age when long-term education was the privilege of a few. A number of men of genius or ability, including some of those mentioned in this narrative, were self-taught of necessity because of impoverished or adverse circumstances. Certainly Green's passion for mathematics would not have allowed ignorance of the French language to stand in his way. Tomlin informs us of his cousin's 'intense application to Mathematics or whatever other acquirements might become necessary thereto': one of those acquirements was certainly the ability to read the works of French mathematicians in the original.

In 1827 Green was thirty-six and a working miller. In scant hours of leisure he had produced a manuscript of text and mathematical calculations which, when printed, would amount to nearly eighty pages. William Tomlin referred to 'Several kind friends [who] . . . assisted him in making known, and in the dedication of his first publication'. Presumably they gave Green the confidence to consider publication. He had two choices: either privately at his own expense, like Goodacre and Toplis, or by submitting his papers to one of the 'learned societies'. By this, Green meant the Royal Societies of London or Edinburgh for which, however, he had no sponsor. Presumably he felt that he did not have much choice and decided to publish privately, since – as he wrote later – the Essay:

> would never have appeared before the public as an independent work if I had then possessed the means of making its contents known in any other way but as I thought it contained something new and feared that coming from an unknown individual it might not be deemed worthy of the notice of a learned society I ventured to publish at my own risk feeling conscious at the same time that this would be attended with certain loss.[27]

This premonition was fulfilled.

The Essay was published in Nottingham in 1828, 'Printed for the Author by T. Wheelhouse'. It has been claimed that only about a hundred copies were printed, though this might be an unreliable estimate.[28] It is, however, true that original copies of the Essay are extremely rare, and can command outstanding prices at auction. In common with the practice of the time, the forthcoming production was advertised in the press, and subscriptions were invited. The advertisement appeared on 14 December 1827 in the *Nottingham Review* – of which Robert Goodacre, son of Green's old schoolmaster, was editor – and in the *Nottingham Journal* the following day:

> In the Press and shortly to be published, by Subscription, An Essay on the Application of Mathematical Analysis to the Theories of Electricity and Magnetism. By George Green. Dedicated (By Permission) to his Grace the Duke of Newcastle, K.G. Price to Subscribers 7s. 6d. The names of subscribers will be received at the Booksellers and at the Library, Bromley House.

The Essay was available by April 1828, and includes a list of fifty-one subscribers. This may not include all purchasers of the book since, by definition, it does not reveal later buyers who could have acquired it from booksellers.

Twenty members of Bromley House Library subscribed to the publication of Green's Essay, and a further six were members of the Bromley House Literary Society, making half the total number of subscribers. A dozen more were Nottingham residents, and there were three from each of the county towns of Lincoln and Derby. Single requests came from as far afield as Ashbourne, Shropshire and Manchester. H.G. Green, with painstaking research, managed to identify all but half a dozen of the fifty-one supporters. This gives some indication of Green's associates. Some would subscribe through personal acquaintance and regard; others, as one later revealed, added their name 'as Country Gentlemen do as a way of encouraging provincial literature',[29] and others simply because patronage was a feature of the age, and the self-respecting burgess or socially aspiring manufacturer conformed to what was expected. It may well be that those who had followed lectures and demonstrations in electricity and magnetism hoped that Green's Essay would provide them with further elucidation. One wonders how many of them, on reading the Essay, remained abreast of his argument, or even if any regretted the expenditure of seven shillings and sixpence for their copy – a sum equal to a poor man's weekly wage. Be that as it may, the list of subscribers provides a significant, if brief, insight into a cross-section of Nottingham's successful townspeople, some of whom Green would have met – if not known intimately – during his years in the Library.

Clergymen, doctors, and the prosperous lace and textile manufacturers and merchants form three groups which made up more than half the subscribers. There were also the dedicatee, the Duke of Newcastle, Sir Edward Bromhead of Lincoln, Lieutenant-Colonel Wildman of Newstead Abbey – who was an old school friend of Lord Byron and the purchaser of his ancestral home – Lady Parkyns of Bunney, the only lady subscriber, and William Strutt, FRS, of Derby. The clergymen included the incumbents of the town's three main churches: the President of the Library, the Reverend Robert White Almond, rector of St Peter's; the Reverend Dr George Wilkins, vicar of St Mary's, the major church of Nottingham, and of St Stephen's, Sneinton; and the Reverend J.B. Stuart, MA, vicar of St James's Church. The Reverend Dr Henry Nicholson was a clergyman-schoolmaster, proprietor of a boarding school in Upper Parliament Street. A number of the medical men who subscribed were associated with the General Hospital, set up by public subscription in 1782, and with the Lunatic Asylum, which had moved to Sneinton in 1812. Of these, the most eminent were Dr Alexander Manson, Fellow of the Royal Society of Edinburgh and a pioneer in the use of iodine in surgery, and Dr Charles Pennington, a founder member of the Library who was physician to the Infirmary, and had a large town practice. Mr Henry Oldknow was a surgeon and his name was well known in the town.[30] No fewer than nine of those engaged in commerce were lace and hosiery manufacturers, reflecting the source of Nottingham's prosperity. Other names – for example, Staveley, Severn and Hind – are still associated with local businesses in or near Nottingham.

Of the individuals among the subscribers known to Green were his cousin and brother-in-law William Tomlin, Robert Goodacre, Samuel Lund, and presumably the President of the Library, the Reverend Robert White Almond, a widely respected figure involved in many of the town's activities. Almond was a Nottingham man, some eight years older than Green; he was educated at the Free Grammar School and one of the few pupils to proceed to university, but not before spending a year being coached by a clergyman in

Lincolnshire. His close friend and companion during this year was the Nottingham poet Henry Kirke White. They both went to Cambridge, Almond entering Queens' College as a sizar, where he was graded Third Senior Optime (i.e. third in order of merit in the Second Class Honours list) in 1808.

When Almond returned to Nottingham in 1814 he took up an appointment at St Peter's, which he held until his death in 1853. He was elected second President of the Library in 1821 on the resignation of Dr Storer, and held the position for thirty-two years. He was a member of the Astronomical Society, had his own telescope, and led scientific discussions in the Debating Society. One of his main duties was to advise on the purchase of new books. His interest in science no doubt accounts for the fact that several scientific societies held their meetings in the Library while he was President: several pieces of scientific apparatus were also fixed in the Library during this time, including the sundial which he presented in 1836.

Almond would certainly have known John Toplis. They were both local men, graduates of Queens' College Cambridge, holding positions of some importance in the town. Although Toplis was some seven years older than Almond, their lives overlapped in Nottingham from 1814 to 1819. Almond may probably be numbered among the 'kind and respected friends' who supported Green, and a representative of the men of culture who not only provided an environment which offered him relief from the cares and restrictions of his family and business life, but also promoted his social development, stimulated his thought, and provided the context in which his genius flourished.

Another subscriber among Green's associates was the Reverend Samuel May Lund, who has already made an appearance as the not very amenable usher of the Free Grammar School. Lund is, as it happens, the only named person with whom it may be claimed that Green had a personal relationship. He was not a member of the Library, but he subscribed to the Essay and he was one of Green's executors. It is unfortunate therefore, that what is known of Lund does not reveal a very agreeable personality. Reference was made above to dissension between Toplis and Wood in the Free Grammar School. It continued between Wood and Lund. The Borough Council, the School's governing body, considered a report[31] in which reference was made to the personal feud between the two men, as a result of which for many months 'they had ceased to hold communication with each other relative to the School or the Business thereof'. In particular Lund faced a charge from parents that 'a system of obtaining attention from the Scholars by the operation of Fear, rather than encouragement, has too much prevailed on the part of the Usher'. He was also charged with compelling boys, as a punishment, to take their books and lessons to his house in West Bridgford, some three miles from the town centre.

Lund had also been appointed chaplain to the county gaol and the House of Correction. The salary for this was £60 per annum, the same as his salary as a – presumably full-time – schoolmaster. New regulations had been imposed earlier, introducing into the school a work period on Wednesday afternoons. Lund protested to the mayor, and enquired 'whether the present regulations be intended to interfere with my duties at the Town Jail; or whether I may be allowed to attend as usual at eleven o'clock on the mornings of Tuesdays and Thursdays. . . . You are perhaps also aware that Wednesday afternoons have always been allowed as holydays. Pray, Sir, is Wednesday afternoon to remain an exception to the general rule . . . ?' If it were, Mr Lund requested additional remuneration for his attendance for four hours on Wednesday afternoon 'holydays'.[32] The crisis point came in 1833. When the elderly Dr Wood retired, the Borough Council determined to get rid of Lund; after some discussion, it was agreed to pay him £100 in order to procure his immediate resignation.[33]

Lund was much the same age as Green, and may have helped him to master sufficient Latin and Greek for his later matriculation at Cambridge. Lund was living in Sneinton in 1839; he had been appointed first curate-in-charge, then vicar, of Awsworth, a village six miles north-west of Nottingham, but in 1838 he was given permission to continue to reside in Sneinton until the completion of the vicarage at Awsworth.[34] One may only conjecture how punctiliously he fulfilled his pastoral care – though absentee incumbents were not unknown – since he is again recorded as living in Sneinton in 1854, in Notintone Place. He was, therefore, a neighbour of Jane Smith and her family. Lund died in Awsworth in 1865 at the age of seventy-four; he is buried in the churchyard there.

Four months after the appearance of the advertisement inviting subscriptions, the Essay was published. Green sent his copy to His

Grace the Duke of Newcastle, with its respectful dedication on the front page. He also despatched a copy to Sir Edward Bromhead with a short accompanying note:

> Sir, I have the honor of sending you a copy of my essay and beg leave to return my sincere thanks for your liberal support.
>
> I remain
>
> Your obliged and grateful servant
>
> George Green
>
> Sneinton – April 19th 1828

Doubtless Lady Parkyns, Lieutenant-Colonel Wildman and others received similar communications. Then George Green waited, and what he had perhaps secretly dreaded came to pass: his private publication had indeed been attended with 'certain loss', for with one exception he appears to have received no response. For what could mathematicians such as Almond, trained twenty years earlier on Newtonian mathematics, make of Green's mathematical analysis? What could the local science enthusiasts understand of his sophisticated development of Poisson's latest memoirs on magnetism? Einstein was to declare that Green, in writing the Essay, was years ahead of his time;[35] certainly young Cambridge graduates of the mid nineteenth century, such as Thomson and Stokes, would avidly seize upon his ideas and develop them; but in provincial Nottingham, in 1828, it was more probably received with polite bewilderment and incomprehension.

George Green Senior was in his seventieth year and probably ailing, for he died the following year, four years after his wife. George was his father's sole executor, and in charge of the family businesses. In July 1829, Jane Smith gave birth to her third child. For the present, therefore, George Green had no time, and perhaps little inclination, to continue his mathematics, and he resigned himself to days taken up with the irksome duty of milling.

Chapter 5

Sir Edward Bromhead

Local subscribers to the Essay would presumably have collected their copies from Bromley House or the bookseller's where they had placed their order. In the case of the nine or ten subscribers further afield and who in all probability were unknown to him, Green would have sent their copies with an accompanying note, curious perhaps as to who they were and how they would judge his work. Each was doubtless sent the formal letter which Green had enclosed with Bromhead's copy of the Essay. Apart from the Duke of Newcastle, Sir Edward Bromhead was the most prestigious of the subscribers. In the light of his mathematical interest at Cambridge and his enthusiasm for the French analysts, Bromhead's subscription to Green's Essay is not surprising. His patronage was ultimately of such support to Green, however, in terms of both his personality and his influence, that it is worth taking a closer look at him.

Sir Edward Bromhead was by now an eminent and – despite frail health – a tireless public figure and benefactor to the city of Lincoln and the county. He was the eldest of the three sons of General Sir Gonville Bromhead, who was created a baronet for his services in quelling the Irish rebellion.[1] Edward Ffrench was born in 1789 and became the second baronet on his father's death in 1822. He died unmarried, whereupon his second brother, Major Sir Edmund Bromhead, succeeded to the baronetcy. The Bromheads are by tradition a military family, but Edward Ffrench and his youngest brother, Charles, neither being of a robust constitution, went to Cambridge. Charles became a Fellow of Trinity College, and later took orders. Edward went first to Glasgow University and then, at nineteen, to Gonville and Caius College, since he claimed to be of 'founder's kin' through his grandmother, Frances Gonville.[2] At Cambridge Bromhead gained a mathematical prize in 1809 and took his BA in 1812. His health did not permit him to undergo the rigours of the Mathematical Tripos, though his ability to do so was never in doubt. In any case, there was no need. Success in the Tripos was essential for those who hoped for an academic career in the University or for entry to the professions. Bromhead's future responsibilities, however, were dictated by birth. A gentleman's education at university and some knowledge of the law, gained at the Inner Temple, were all that was necessary for him to discharge them.

Bromhead became much involved in city and county affairs. He was High Steward of Lincoln[3] and by 1828 he had drawn up 'Rules for the Government of the County Gaol and Castle of Lincoln', which he sent as a 'Present' to the Royal Society of London of which he became a Fellow in 1817.[4] In 1816 he was a committee member of the Lincoln Library, founded two years earlier; in 1822 he was the first President of the new Permanent Library founded to serve the lower social class of tradesmen, mechanics and apprentices. As chairman of the first inaugural meeting, he established the non-political nature of the enterprise: 'here we are, whigs, tories and I believe, radicals, all agreeing and combining to set the [River] Witham on fire'. Bromhead was also approached by local ministers to help to establish a Temperance Society;[5] he had constructive suggestions for the provision of coffee, cider and ginger beer in place of alcoholic drinks.[6] It was possibly he who contributed the judicious ruling that 'brandy in puddings was eating, not drinking, and therefore not under ban'. Sub-Dean Bayley[7] appealed to him for help in drafting the rules of a proposed Friendly Society, and in 1838 he was asked to chair a meeting for the selection of a parliamentary candidate. He advised on land tax and reform, and was interested in projects to start steam navigation on the River Witham. As his friend Dr Charlesworth wrote in 1827, when Bromhead was thirty-eight: 'Nearly all the institutions of Lincoln have already the stamp of your mind and hand: several have been entirely formed by you, and others totally re-modelled . . .'.[8]

Bromhead's influence and expertise were recognized in less provincial circles. In addition to his Cambridge contacts he kept in touch with friends in London, travelling there when his health allowed. He was elected a member of the Royal Society of Edinburgh in 1823. In 1825 he was approached by Robert Peel. Peel, later Prime Minister, was then Home Secretary, and engaged in revising and enacting laws on many urgent problems, penal, agricultural and social. Peel had the reputation of arriving in the Commons with all the information his bills required, culled from expert sources. Bromhead's letter may have been such a source:[9]

Thurlby Hall
Lincoln

Sir Edward Ffrench Bromhead presents his compliments to Mr Peele [*sic*], is willing to employ his leisure until the next meeting of Parliament in proposing a Bill on one of the proposed subjects.

'A Bill for explaining, ammending [*sic*] and reducing into one Act of Parliament, the laws for the repair of the Highways not being Turnpike'.
'A Bill for explaining, ammending and reducing into one Act of Parliament, the Laws for the Relief of the Poor, except as far as relates to Houses of Industry'.
'A Bill for the more effectual Regulation of Summary Proceedings before Justices of the Peace and others'.
Before undertaking so serious a task, he would wish some assurance, that the Bill [which] shall be brought before Parliament is drawn in a satisfactory manner.
July 9th 1825.
The Right Honourable Sir Robert Peele
etc. etc. etc.

This was but one aspect of Bromhead's activities, as magistrate and public administrator, but his correspondence also indicates his continued interest in mathematics. One letter, dated 1 May 1818, is to the editor of the *Encyclopaedia Britannica*:[10]

Thurlby Hall
Newark
Notts
May 1st 1818

Sir,

I shall have much pleasure in writing the article '*Differential Calculus*' for your supplement to the Encyclopaedia Britannica. The article will not be long, and will be wholly distinct from the valuable Memoir on *Fluxions* in the original work, consisting chiefly of new views and original theorems. For my ability to execute such an undertaking, I must refer you to the Transactions of the Royal Society of which I have the honour to be a member. Should my proposal meet with your wishes, I would request to know the time at which the communications will be wanted. I have the honour to be Sir

Your very obedient Servant
Edward Ffrench Bromhead

Bromhead duly wrote the article, which was published in the Supplement to the *Encyclopaedia Britannica* of 1819. Reference to names such as Woodhouse, Herschel, Babbage, Whewell and Peacock are an indication of Bromhead's approach to his account of the differential calculus. Three years earlier he had sent to the Royal Society his paper on 'Fluents of Irrational Functions', which was published in the *Transactions*. Bromhead's expertise was maintained in the liveliest fashion through his correspondence with his lifelong friend Charles Babbage. In the letters are pages of mathematical calculations, and problems are passed from one to the other. Other issues are discussed, including Babbage's plans for his calculating machines. 'Whatever else you do, I hope you will finish your Memoir describing the Mechanism of your Instrument, there will be immortality in that, whether you finish it or not', Bromhead wrote in July 1823,[11] with greater truth and vision than he could ever have imagined. Earlier he had advised Babbage on his proposal to form an Astronomical Society: 'I think it would be an improvement to enact that no person should be president for more than three years in succession. This would keep it constantly alive as every president would want to make a dash during his office.'[12]

Bromhead maintained his contacts with Cambridge in order to attend Convocation and election meetings, browse in the bookshops, and see his younger brother Charles. William Whewell, like Charles, was a Fellow of Trinity, and it was probably through Charles that Bromhead got to know him, since their undergraduate course overlapped by only a year and Whewell was not a member of the Analytical Society. Whewell became an influential figure in Cambridge, and was later instrumental in getting Green's work published there. He was Master of Trinity from 1841 until his death in 1866. He was a prolific writer on many topics, the author of a number of university textbooks, and the subject of many anecdotes,[13] which tend to stress his omniscience, the firm authority wielded by the Master in his College – and less happily – his strong sense of self-importance.[14]

Bromhead never married, but this did not deprive him of a life of personal relationships and family affection. The health of his brother Charles in the 1820s was a cause of much concern, as was the care of his widowed mother, who 'is now with me and in such delicate health that I should not feel justified in leaving her', as he wrote to Babbage in 1823.[15] His letters reveal both wit and humour. He was touched by Babbage's desire for him to be godfather to his son (Babbage's eldest son had been given the name Herschel, since the two men were close friends). Bromhead was happy that his Christian name would be chosen, for 'indeed there is a dreadful want of euphony in "Bromhead Babbage".'[16] 'Love to the Babbagelings and my respectful compliments to Mrs. Babbage' concludes another letter.[17] Bromhead was also a keen botanist. He may have sponsored one of the various expeditions of the time to discover new and exotic species of plant, since an 'orchid bromheadia' has been found listed in a catalogue.[18]

In 1842–3 Bromhead spent £1000 on the restoration of Thurlby Church,[19] replacing the thatched roof with slates, paving the earth floor and installing new pews and choir stalls. With an exactitude which may reflect his attitude to wider concerns, he recorded on the inside of the church door: '1843 This door of Thurlby oak is a copy of the old door of ash except that the old door was not grooved and had only two hooks. E.FF.B.' These initials, also found elsewhere, are the only memorial to Edward Ffrench Bromhead in the church he restored, in which so many wall plaques and windows commemorate his forebears and successors. He eventually went blind, and died at the age of sixty-six. One may look almost in vain for his grave in a churchyard where Bromheads of many generations are buried, since Edward Ffrench requested a simple cottager's stone. He lies, as he requested, facing the west wall of the church, the foot-high tombstone incorporated into an old brick wall overgrown with foliage, the inscription now barely legible.

Bromhead's life might be better known had he entered more

into national rather than local affairs. His enquiring mind and diversity of interests may well have been a disadvantage in that his talents were not focused in any clearly defined direction. More particularly, perhaps, he was handicapped by the lack of a strong constitution – which paradoxically, given the family tradition, might have established him in an army career, emulating a Bromhead who fought with Wolfe at Quebec, or another with Wellington at Waterloo, or a third who won a VC at Rorke's Drift in 1879.[20] An occasional letter refers to his condition. He finds that he cannot undertake his proposed journey to London, the pain in his chest inhibits all movement. 'To write at all is an exertion' he says in a four-line letter to Babbage in 1829.[21] And then finally, failing sight brought his life to a close in darkness in 1855.

Edward Ffrench's kindness, his humour, his lively mind and his concern for others – not least his patronage of Green – offer a portrait of the English country gentleman of his time. Yet Bromhead, as a later letter will reveal, was a man of his period, duly conscious, beneath his ease and charm of manner, of social distinctions. And George Green likewise, as his letters will reveal, was also conscious of the social distinctions of the time.

Green's note sent with the Essay was dated 19 April 1828. Left at the departure point of one of the coaches to Lincoln, the package swiftly arrived at Thurlby Hall, since on the following day, 20 April, Bromhead sat down to write a letter of acknowledgement. This letter is unique in that it is the only one of Bromhead's letters to Green to have come to light. All the others have disappeared with the rest of Green's papers; as a result, the correspondence between the two men is tantalizingly one-sided. The letter is also unusual in that it is one of the very few drafts of letters retained by Bromhead amongst his correspondence. This is significant in that it reveals the usually confident and articulate Bromhead carefully feeling his way in an unknown and delicate situation – hence the hesitations and erasures. He knew nothing of the writer of this remarkable work (which he must immediately have read, if not closely studied). The offer he was making might prove patronizing to one of his own class, yet how was he to rescue and encourage a mathematician of such promise and originality?[22]

April 20, 1828

Sir

I have had the pleasure to receive your obliging note, ~~and~~ with ~~Memoir~~ the Work on Electrical Attractions and Magnetic Forces – You will not I hope consider me as taking a Liberty, when I ~~offer~~ ~~express~~ ~~my~~ say that I shall feel ~~xxx~~ happy in communicating ~~xxx~~ any Memoir which you may hereafter compose, (to the Royal Societies of London or Edinburgh –) In this Country the number of scientific Readers is so extremely limited, that ~~none~~ ~~would~~ ~~be~~ elementary works alone can admit of an independent Publication – All Memoirs of a profound Nature must look for support and perusal to the circumstance of appearing in some Volume of Transactions or some Encyclopaedia. If these Remarks should unfortunately apply to your present Publications, I sincerely hope that you will not feel discouraged from pursuits in which you ~~can~~ ~~will~~ may certainly attain a very distinguished Reputation –

I remain ~~and~~ ~~am~~
Sir
Your very faithful Servant

Mr. George Green
Sneinton
Nottingham

At the same time, Bromhead set an enquiry in motion to find out more about Green from a source in Nottingham; the answer provides a unique contemporary bird's eye view of the mathematician.[23]

1828

Sir

I learn from Nottingham that Mr G Green is the Son of a Miller, who has had only a common education in the Town, but has been ever since his mind could appreciate the value of learning immoderately fond of Mathematical pursuits, and which attainments have been acquired wholly by his own perseverance unassisted by any Tutor or Preceptor: he is now only 26 or 27 years of age of rather reserved habits attends the business of the Mill, but yet finds time for his favorite [*sic*] Mathematical reading –

Your obt. Servant
Thos Fisher

Asylum May 10

Thomas Fisher was the Director and Secretary of the Lincoln Asylum, of which Bromhead had become a Trustee the previous year. Fisher's connection with Nottingham is unclear. He was not a member of Bromley House Library, but he had subscribed to the Essay. He may have been related to John Fisher, a baker in Carter Gate, who would have known George Green Senior, a fellow baker and miller – particularly as Carter Gate was on the east side of Nottingham, scarcely a mile from Sneinton. Fisher's letter, too, is unique, since there is no other contemporary report on Green's appearance and manner. William Tomlin's formal account was written in 1845, four years after Green's death, and retails events of many years earlier. Fisher's short note describes Green as he appeared to neighbours and acquaintances who were in no position to know much about his private life. There are points in common with Tomlin's account: Green's passion for mathematics, the fact that he was (or appeared to be) self-taught, and the information that he was much involved with work at the mill. What is intriguing is the comment that 'he is now only 26 or 27 years of age', since he was in fact already thirty-five.

Bromhead, perhaps to his surprise and disappointment, did not receive a reply to his letter to Green until twenty months later, when the following lengthy missive arrived at Thurlby:[24]

Sneinton near Nottingham Jan 19th 1830

Sir

From some observations made to me last Saturday by Mr Kidd of Lincoln I find that I have unintentionally been guilty of a gross neglect on an occasion where of all others I would most carefully have avoided it and therefore hope you will pardon the liberty I am about to take in endeavouring to explain the circumstance of my not having answered your very obliging and condescending letter and this explanation I am the more desirous to enter into because nothing connected with the publication of my little Essay has afforded me so much satisfaction as that it should have been found in any degree worthy of your notice.

Had I followed my own inclination I should immediately have

written in order to have expressed in some measure my gratitude for the very handsome offer with which you had honored me but on mentioning my intentions to a gentleman on whose opinion I had at an early age been accustomed to rely he assured me that no answer would be expected but that on the contrary it would be considered as a liberty to trouble one so much my superior farther until I should be able to avail myself of your kind offer by forwarding some memoir to be communicated to one of the Royal Societies and as this gentleman had seen more of the world than myself I yielded to his opinion though with reluctance lamenting at the same time that custom should compell me to act in a way so much at variance with my own feelings.

What has just been advanced necessarily places me in rather a ridiculous light I was however determined for my own satisfaction to make a frank confession of what had taken place and trust entirely to your goodness to excuse an error arising from little knowledge of the world and consequence of having devoted almost exclusively to books the few leisure hours I have been able to snatch from the tedious and uninstructive details of common business.

Although from a mistaken notion of propriety I have been so long hindered from making any acknowledgement for the very handsome offer you were so kind as to make I trust you will do me the justice to believe that I have felt most sensibly the honor conferred upon me by so much condescension on your part and that I have always esteemed that offer as most valuable. Indeed the trifle which you honored with your notice would never have appeared before the public as an independant work if I had then possessed the means of making its contents known in any other way but as I thought it contained something new and feared that coming from an unknown individual it might not be deemed worthy of the notice of a learned society I ventured to publish it at my own risk feeling conscious at the same time that this would be attended with a certain loss.

Had it not been for a severe domestic affliction which has thrown me into a situation entirely new and so prevented me from paying the least attention to mathematical subjects for the last twelve months I should long since have finished a little memoir somewhat analogous to my Essay with this I had intended troubling you and indeed if I could flatter myself that you would forgive the awkward blunder of which I have been guilty I would make every effort to complete it although the time necessary for doing so were stolen from my sleep.

With the deepest sense of gratitude for the favors already conferred.

I remain
Yours Most Respectfully
George Green

The mystery concerning the identity of the misguided gentleman who discouraged Green from responding to Bromhead's offer remains; nor is there information on the identity of the helpful Mr Kidd, except that he was a subscriber to the Essay. In speculation about the former, Goodacre, Toplis and Almond have been suggested,[25] but Toplis had left Nottingham nine years earlier, and Almond had arrived only in 1814, when Green was twenty-one. Robert Goodacre Senior would appear to be the most likely candidate, since as Green's schoolmaster he had certainly known him 'from an early age'. Goodacre had travelled the country giving lectures on scientific topics, and his itinerary could well have included Lincoln, where he might have heard of Bromhead, or even met him, since Bromhead would certainly have lent his patronage to such an event. The obituary of Robert Goodacre Junior stated that he was also at one time clerk to the Lincoln Board of Guardians, which suggests a possible second connection between Lincoln and the Goodacres.

Bromhead did not leave Green waiting for a reply to his long and apologetic letter, since on 13 February Green had written a second letter,[26] explaining that he had delayed acknowledgement of Bromhead's reply in order first to read the memoir Bromhead had sent him: 'I can assure you I have derived both pleasure and instruction from its perusal.' The work, he adds somewhat pedantically, in dealing with a difficult branch of the integral calculus, 'had furnished luminous and scientific views for the direction of the mind in some laborious researches in which it had formerly been left to grope, as it were, unaided in the dark'. More significantly, he added: 'in consequence of the encouragement contained in this letter I have to a certain extent recommenced my mathematical pursuits and trust that before very long I shall be able to draw up a little paper which probably would never have been effected but for your kindness and condescension'. Bromhead's memoir, incidentally, was almost certainly his own paper on the 'Fluents of Irrational Functions'.

Green's next letter to Bromhead is dated over two years later than the one just quoted: 17 May 1832.[27] The long interval between the two letters would indicate that during this time Green paid several visits to Thurlby, a village halfway between Newark and Lincoln, and two miles east of the Foss Way (A46). Coach travel from Nottingham was considerably developed by that time. Five coaches left daily for London and two for Cambridge; three left each day for Newark and Lincoln: at 5.30 a.m., 8.30 a.m. and 4.00 p.m.[28] Roads were considered good for the period, better than those within the town boundary. An alternative to taking the stagecoach in summer would have been to travel on horseback. Green's uncle and cousins still farmed at Saxondale, little more than a mile from the junction of the old London Road (A52) and the Fosse. As horses were then a normal means of transport and there was ample stabling at the mill, Green could have ridden from Sneinton to Stoke Bardolph, and thence gone by ferry across the Trent to Shelford and Saxondale.[29] From there it is eleven miles to Newark and a further twelve miles or so to Thurlby, a journey of about two hours on horseback. On his death, George Green left an annuity of ten pounds per annum to his uncle, Robert Green, who lived in Saxondale.[30] This suggests a continued association with the family, so Green may have broken his journey to Thurlby and rested his horse at the family farm.

The village of Thurlby, now only a hamlet, comprises the Hall, the church and a dozen houses. The small church, with Edward Ffrench's oak door and the poppyhead pews, stands back from the lane behind its surrounding trees; its grassy churchyard is quite large and utterly peaceful. In Bromhead's day, the village of about one hundred and forty-five inhabitants[31] would have seen more coming and going than it sees now. The Hall was the pivot of village life, the centre of patronage, employment and charity, as was the 'Big House' in hundreds of villages in the rural England of the eighteenth and nineteenth centuries. Members of the Bromhead family came and went – Edmund to his regiment, Charles to Cambridge. Edward Ffrench was always busy, with frequent journeys to Lincoln, to Sleaford and elsewhere in the county, interspersed with visits to the Babbage family, to Cambridge and, when possible, to London. And doubtless visitors came to Thurlby – we have seen that Bromhead presided over the gathering of members of the Analytical Society at the Hall in 1817.

One visitor will have arrived with mixed feelings, possibly in some trepidation, having but 'little knowledge of the world', as Green had confessed in his first letter to Bromhead. But the latter, only four years older than Green, with his long experience of men and affairs, would have known how to set the miller at his ease, and once the subject of mathematics was broached, social inequalities would for the moment have been forgotten. Yet the world of Green and Bromhead was one of rigid social structure, as is evident in the letter Bromhead wrote in 1845, four years after Green's death.[32] It was sent to the Reverend J.J. Smith, Tutor at Gonville and Caius College, when enquiries were being made about Green, and it is valuable in giving Bromhead's account of these meetings at Thurlby.

Lincoln March 24 – 1845

My Dear Sir

My acquaintance with the late Mr. Green was quite casual. I met with a subscription list for his first mathem: publication, and added my name as Country Gentlemen often do by way of encouraging every attempt at provincial literature. When the work reached me it was obvious from the limited number of subscribers and their names that the publication must be a complete failure and dead born. On meeting with a person who knew him, I sent him word that I should be glad to see him and he visited me accordingly at Thurlby –

We of course had a great deal of scientific conversation, and he told me that he found so little sympathy with his pursuits, that he had finally relinquished them. Against this I warmly and successfully remonstrated (as i [*sic*] have done in the like case of another distinguished Mathematician[33]) explaining to him that such a publication as his not being elementary or systematic could not stand alone even in London and that against the provincial press there was moreover a regularly organized conspiracy.

He mentioned to me some original views which he had, and I

> persuaded him to work them into the form of a memoir, which I undertook to lay for him before the Royal Societies of London or Edinburgh or our society at Cambridge by which he would be saved from much trouble and expense and would find his views fairly brought before the scientific public and estimated at their value . . .'

Green must have returned to Nottingham after this meeting remotivated and re-enthused by the 'scientific conversation' and Bromhead's assurance of help in further publications. As we have seen, his father's death the previous year had not only released him from his filial duties at the mill – his inheritance now allowed him to live as a 'gentleman', with time to follow his mathematical pursuits. Green therefore leased the mill[34] and sold the business. These transactions, together with the rents from his father's Nottingham property, released him from daily toil, and he could now devote his time to mathematics. Bromhead's encouragement soon bore fruit.

Chapter 6

The Publication of George Green's Further Investigations

A significant event for those interested in the life of George Green was the discovery, only some twenty years ago, of his letters to Bromhead, written between 1830 and 1834.[1] These represent a very considerable bonus – not only because of their rarity, but also because they give a unique insight into Green's thoughts and problems in mathematical matters, and offer a very rare glimpse of his personality. If one half of this unique correspondence had to be lost, it is fortunate that it is Green's letters that have survived. It is frustrating, however, that none of his working papers is extant, so it is not possible to see the detailed development of his ideas, or the false starts and the culs-de-sac in which he may have found himself. The result is that on occasion Green appears to make assumptions or proceed to generalizations where it would be helpful to know how he arrived at his conclusions. As it is, the impact of the Essay, coming out of the blue, is staggering. It is not surprising that Bromhead responded as he did, offering to make public any further work from the same mind and pen.

Green employed the next two years working on a second investigation, then on 17 May 1832 he wrote to Bromhead:[2]

> I fear you will think my paper rather too hypothetical but as all the results therein contained may be applied to the electric fluid provided we give to the general exponent n the particular value 2 and as moreover every successful attempt to determine mathematically how an infinite number of free particles will comport themselves supposing them to act upon each other

mutually with forces varying according to a given law appears to belong to a class of mathematical investigations more likely to throw light on the intimate constitution of natural bodies than many others. I hope it will not be found entirely destitute of interest. Being quite at a loss for a proper title I have ventured to propose the following –

'Mathematical Investigations concerning the Laws of the Equilibrium of a Fluid analogous to the Electric Fluid together with other similar researches.'

Though I had originally intended my Paper for the Royal Society in London I shall leave the disposal of it entirely to your better judgement and shall feel particularly obliged if you would inform me to whom it will be necessary to forward a copy of my Essay in case you consider such a step advisable.

I cannot conclude without expressing my gratitude for your kind assistance which can only have arisen from a liberal desire to forward the interests of Science by encouraging even the most humble cultivators.

I remain with the greatest respect
Yours Very Sincerely
Geo : Green

Bromhead duly forwarded both the memoir and the Essay to his friend Whewell for reading by the Cambridge Philosophical Society. Unfortunately, Green's memoir, sent presumably in late May or early June, arrived just as the College was dispersing for the long vacation. 'The memoir you sent me', wrote Whewell to Bromhead on 14 November 1832, 'arrived as our Society was breaking up for the long vacation and I have not had an opportunity till now of laying it before our mathematicians.'[3] Hence Green learnt of its reception only some six months after its despatch. It would be another four months before his work would be in print. The delay must have seemed long to Green, as Bromhead understandingly wrote to Whewell on 1 April 1833: 'The appearance of this production must be as interesting to Mr Green, as a debut on the Stage, a Maiden Speech, or the first appearance at a Ball . . .'.[4] In actual fact, the final publication was not all that protracted. The procedure was that the paper was 'read' – that is, given approval by the two moderators who judged it worthy of publication. In terms of attribution of its contents this date could be important, since two or three years could pass before space could be found for it in the annual *Transactions*. In Green's case, the paper was read on 12 November 1832 and printed the following year. The full title of its entry therefore reads:

> 'Mathematical Investigations concerning the Laws of the Equilibrium of Fluids analagous [*sic*] to the Electrical Fluid, with other similar Researches. By George Green Esq. Communicated by Sir Edward Ffrench Bromhead, Bart., M.A., F.R.S.L. and E. (Cambridge Philosophical Society, read 12 November 1832, printed in the *Transactions* 1833. Quarto, 63 pages)'

Work had to be done on the memoir before it could be published, but Bromhead and Green wasted no time. Whewell wrote to Bromhead in his November letter that their mathematicians had found it 'a very good piece of analysis with some new and curious artifices employed in the reasoning', but advised some curtailment if the author agreed. A week later Green was writing to Bromhead agreeing fully to some abbreviation and requesting guidance. 'If you think by going to Cambridge I should be better able to do this, I would readily undertake the journey . . .'.[5] Bromhead, however, writes to Whewell:

> Would it be too great a favor to request that you would become Gardener to this pruning . . . Mr Green had retired in Despair from Mathematics, and undertook this Memoir at my request, from which you will see that a little encouragement may secure him as a Recruit to the very small troop who serve under the severe Sciences . . .'

Accordingly, Whewell returned the memoir with notes on its proposed abbreviation: rewriting it would, as Bromhead pointed out, entail a second reading. The rapporteur was Robert Murphy, Fellow of Caius College and someone who will inadvertently play a small but significant part affecting Green's posthumous reputation.[6] Murphy sent a copy of his own paper on definite integrals for Green's perusal. By 5 January 1833, Green told Bromhead that he could suppress six or seven sheets and curtail others: this had taken more time than he expected, but he hoped to finish by the following

Monday. 'If you could spare an hour on any following day, I would wait upon you in order to have your opinion on the propriety of the alterations which have been made.'[7]

Bromhead duly sent the revised memoir to Whewell,[8] and asked: 'If the Society allows extra copies to be run off, we would take advantage of the Indulgence'. By February the proofs were ready, but Green was happy to follow Bromhead's advice for Professor Whewell to give them a 'cursory review' to avoid the delay incurred in sending them to Nottingham. On 1 April Bromhead reported to Whewell Mr Green's gratitude 'for the handsome presswork of his memoir', and by 5 April Whewell was forwarding forty copies to Bromhead. 'The Society gives each author 25 copies', but he was sending twenty-five additional copies, 'the expense of this to Mr. Green, including everything to the stitching of the copies will be less than thirty shillings'. The publication in the *Transactions* would take place in a month or two; meanwhile, he was distributing ten copies to 'Some of our best mathematicians here'. Green was, one assumes, very pleased. 'It is into the hands of such men that I should wish them to fall', he wrote to Bromhead, who finally forwarded Green's payment of costs to Whewell on 28 April 1833.

Green, now apparently in full spate of intellectual activity, was already at work on two other investigations, and the writing and disposal of these are also subjects for discussion in his letters to Bromhead. Thus in the previous February,[9] when he agreed to the proofreading in Cambridge, Green intimated that he had material in hand for a paper to be sent 'into the North'; the following month he mentions yet a third paper (he calls it his second memoir), which:

> is now almost finished and unless I deceive myself will be found more worthy of the notice of the Cambridge Society than the preceding one. However as I may be thought unreasonable in troubling them with a second paper so soon after the one which they intend honoring with a place in their Transactions and as I am naturally anxious to bring it forward before the long Midsummer Vacation if anything should occur to you as to the proper disposal thereof I should be very thankful if you would inform me of it. Of course I should prefer sending it to Cambridge if it would be likely to meet with a favorable

reception there.

Bromhead had already alerted Whewell to the possibility of a second paper, which – he suggested – would not have the prolixity of the first: 'The next contribution will be condensed and carefully written, and I have reason to hope of some value.'[10] Green completed his memoir 'On the Determination of the Exterior and Interior Attractions of Ellipsoids of Variable Densities' by 27 April and sent it to Bromhead, accompanied by a long letter:

> I now take the liberty of forwarding my new memoir, which, as is usually the case, has taken more time in arranging than was first calculated upon; and tho' I have endeavoured to condense, as much as possible, it has attained a considerable size, greater, probably, than you will think the subject entitled to: I shall therefore trouble you with a few words in explanation.
>
> You well know that the problem of determining the attractions of ellipsoids, was, at one time, considered the reproach of analysis, and I think, it must be confessed, that so long as mathematicians confined themselves solely to ingeneous transformations of the original integrals their success was not commensurate to the time and talents which were brought to bear upon it. The method that I have employed is entirely different, and, by dispensing with the labor of the integrations, enables us to give under a simple form the attractions of the ellipsoids, when the density, instead of being constant, is expressed by very general functions. Indeed I flatter myself that the simplicity of the results is such as to place in a clear point of view the great advantage of the modern methods, even where they had formerly been thought of less avail, and thus to remove the reproach before mentioned.
>
> Though I imagine that to have resolved a celebrated problem on far more general hypothesis than had ever yet been attempted may serve, in some measure, to excuse the length of the paper, it is not on this plea that I chiefly rely: For, unless I much deceive myself, it will be found to contain methods of considerable generality capable of many interesting applications, if, therefore, you should conceive a case has been made out, sufficiently

> good to entitle it to go to Cambridge, I should feel much obliged to you to forward it; but, if otherwise, would you be kind enough to choose some lower and more suitable destiny?

Bromhead despatched it to Whewell the following day,[11] trusting, no doubt, that it could be read before the long vacation.

> I send another memoir, rather unwillingly, considering the late trespass on the Transactions, but the memoir seems so very new and important that I dare not suppress it, or send it to London or Edinburgh, without giving you the refusal, as we Lindites put it – The Cambridge Transactions ought to lead all others in Mathematics . . .

Green, having waited six months for news of his first memoir, wrote to Bromhead within a month of the despatch of the second:

> Should you hear anything about my paper on the Attractions of Ellipsoids be kind enough to inform me of it for to confess the truth I consider this decidedly superior to the former and feel tolerably confident that it will effect without flaw what it professes. With this belief I shall be the more anxious to know whether or not other and more impartial judges think the same.[12]

The thirty-five-page memoir on ellipsoids was read on 6 May 1833, though it was not published in the *Transactions* until 1835.[13]

Green now had two papers accepted by the Cambridge Philosophical Society. He was working at the same time on the third paper, which Bromhead was proposing to send to Edinburgh. Bromhead obviously felt that he could not offer it to Cambridge, since he had proffered the second memoir on ellipsoids with some diffidence. The Royal Society of Edinburgh was, in Bromhead's view, the obvious alternative, since the Royal Society of London was not to be considered at this time. Bromhead and his friends had sent papers for publication there in the past, but since the foundation of the Cambridge Society in 1820 they had considered this the appropriate outlet for the publication of new works. Only three years before, in 1830, Charles Babbage had published his celebrated attack on the Royal Society in

Reflections on the Decline of Science in England, which was in principle a denunciation of both its organization and its scientific standing. Green, on his own admission, deferred entirely to Bromhead in the matter of the publication of his work, so none of his papers had what would have been the valuable posthumous distinction of publication in the *Transactions of the Royal Society*.

In February 1833 Green wrote to Bromhead:

> Should you think it advisable for me to send a paper into the North I have now the material for forming one on the determination of the correction to be applied to the observed times of vibration of a pendulum moving through exceedingly small arcs in a fluid of given density to have the same time in vacuo. When the body of the pendulum is of an ellipsoidal figure situated in any way with regard to the plane of its motion this correction is readily deducible from the general theory of the motion of fluids as contained in the Mecan Analyt.[14] and other works but I find from what is contained in the first paragraph of a memoir in the London Transactions for 1829 near the end of one of the parts that it is not unlikely but M Bessel[15] may have anticipated me. The theoretical researches of M Bessel are contained in the 128 No of Prof. Schumacher's Astronomische Nachrichten Jan 1828. If you could give me any information of their nature it would be conferring a very great favour and probably would enable me to determine whether I have really been anticipated or not.[16]

The article containing the reference was written by Colonel Edward Sabine of the Royal Artillery, who was also Secretary of the Royal Society; it was entitled 'On the reduction to a vacuum of the vibrations of an Invariable Pendulum'. This query seems to have been answered by Bromhead, and three months later Green's paper 'Researches on the Vibration of Pendulums in Fluid Media' was sent to Thurlby; 'it has been so hastily written that I fear you will not consider it entitled to go to Edinburgh'. Green was now alert to the advantage of conciseness of style. 'I take the introductory sheet to be the least and have marked it with a (o) which' – he adds, with a rare touch of humour –

> will likewise serve in my opinion to represent its value very nearly. . . . By omitting sheet (o) the memoir would certainly become shorter and probably on that account better. Indeed had I not thought that the importance of being acquainted with every circumstance which can affect on any supposition the motions of pendulums might serve to give an interest to investigations in other respects trivial enough I should hardly have run the risk of forwarding the present communication. I have however some confidence in the results since they agree with others found by a very different method and contained in a 'rudis indigestaque moles'[17] of papers relative to the motions of fluids written many years ago at such intervals of leisure as I could snatch from other avocations when I had less spare time and greater industry than at present.[18]

Green's memoir on the vibration of pendulums was read on 16 December 1833, but it was not published in the Edinburgh *Transactions* until 1836. The acceptance came in a letter to Bromhead dated 1 June from the President, Sir David Brewster: '. . . there can be no doubt that any paper written by so excellent a mathematician, and communicated by you, will find a place in the Edinburgh Transactions'.[19]

Hydrodynamics was a more widely discussed topic at this time than electricity or magnetism, and this was the first of three papers on the subject that Green would eventually write. Green's reference to previous work 'many years ago' is a reminder of what has been lost in the disappearance of his working papers, since it would have been instructive to compare the earlier, 'very different method' with that employed in the completed memoir. Green had already given an indication of an earlier interest in hydrodynamics in his letter of 27 April which accompanied his paper on ellipsoids:

> By the last paragraph of the introduction you will perceive, that my analysis might readily be applied to certain cases of the Theory of the motion of Heat, and likewise to the laws of the Motion of Fluids; two subjects that you were kind enough to recommend to my notice. In a particular case of the latter I could readily draw up a memoir which would, at least, have the merit of being very short.[20]

Bromhead's suggestions of subject may have been prompted by

similar suggestions made by Whewell in his letter of 5 April, a fortnight earlier, when he had sent the forty copies of Green's memoir on the equilibrium of fluids:

> The publication of the part of our Transactions in which this memoir will appear will take place in a month or two. I have no doubt that any memoir of equal merit with the last would be favourably received by the Society. Still I do not know that we should be in *haste* to print such a paper: the printing is expensive and we wish to vary our transactions which are apt to contain too much of pure mathematics. You will observe I speak with the caution becoming an official person and by no means wish to deter Mr Green from sending us something more.

Whewell then proceeds to suggest probable fields of investigation, but first he urges a regard for realistic study:

> For my own part I may observe that I read with more pleasure those analytical investigations which apply to problems that really *require* to be solved for the purpose of advancing physical science . . . than to those researches where the analytical beauty and skill are the only *obvious* merits.

Thus:

> extension and simplification of work on the laws of heat (by Fourrier) [*sic*], on the motions of fluids (Poisson, Cauchy) both with respect to the properties of waves and the laws of the tides (Laplace), on the many problems the theory of light offers . . . may supply ample employment for the most refined analysis without deviating from the great patch of mathematico-physical investigation.[21]

The titles of Green's subsequent investigations suggest that he took Whewell's advice.

Meanwhile, Green was still pursuing researches on hydrodynamics stemming from the Edinburgh memoir. He opens his letter of 8 June 1833 with the fullest expression of gratitude to Bromhead for his help and advice, and then proceeds once again to seek Bromhead's help in clarification of a point he has come across in his reading:

> In returning Sir David's letter I avail myself with pleasure of the

> opportunity thus afforded to assure you how grateful I feel for the trouble you have taken on my account and the kind advice which you have always so generously offered. The success of my memoirs has been much greater than I anticipated and is I am persuaded principally due to the sanction of your name and to having strictly followed your advice as to the proper disposing of them. I hope my next paper will convince you that I have not neglected it in the choice of a subject.

He has read Herschel's *Treatise on Astronomy* in Lardner's *Cyclopaedia*,[22] in which there is a reference to William Whewell's investigations relative to the tides. There may be a 'little corner' in the General Theory of the Motions of Fluids still open . . . 'If on further investigation this prove correct I shall endeavour to gain possession of it.'[23] Whewell was in fact conducting investigations into the phenomenon of the tides, and published in that year the first of fourteen memoirs on the subject in the *Transactions of the Royal Society*. Presumably Bromhead was able to settle the point, but the final result in the form of a memoir 'On the Motion of Waves in a Variable Canal of Small Width and Depth' had to await the end of Green's degree course and was not published in the Cambridge *Transactions* until 1838.

Both William Tomlin and Edward Bromhead recounted in their letters of 1845 that Green had been encouraged to think of going to Cambridge 'several years before that circumstance actually took place' – Bromhead suggested that the subject had been discussed early in 1830. In April 1833, Green himself raised the matter in a letter to Bromhead: '. . . you are aware that I have an inclination for Cambridge if there was a fair prospect of success. Unfortunately, I possess little Latin, less Greek, have seen too many winters, and am thus held in a state of suspense by counteracting motives.'[24] Bromhead, with his customary kindness, proposed that Green should accompany him to Cambridge to meet his friends there. Herschel was about to depart for the Cape of Good Hope to spend two years plotting the constellations of the southern hemisphere. The gathering on 24 June of Bromhead, Herschel, Babbage and others (Airy, perhaps? Sedgwick? Hopkins?) was thus a reunion of old College friends and a farewell party of goodwill. Understandably, an outsider such as Green might well have felt *de trop*. Inevitably, for one of his reticence, independence and social background, it would have been too much of an ordeal. Writing in May, he refused the invitation:

> You were kind enough to mention a journey to Cambridge on 24 June to see your friends Herschell Babbage and others who constitute the Chivalry of British Science. Being as yet only a beginner I think I have no right to go there and must defer that pleasure until I shall have become tolerably respectable as a man of science should that day ever arrive.[25]

Were he to arrive in Cambridge, it would be as a forty year old undergraduate: these men, his contemporaries in age, were already College Fellows, University Professors and Fellows of the Royal Society, with publications to their name and a growing reputation. None the less, in his next letter – dated 8 June, the last he would write from Nottingham – Green had made the decision to go to Cambridge:[26]

> Under an impression that it would be worth while to spend some years in Cambridge I have let the house which I now occupy so that I cannot remain strictly in my present locality but can nevertheless situate myself so as to be completely beyond the reach of the temptations to mental dissipation which you mention though were I actually in Cambridge I flatter myself I could resist them having a tolerable share of obstinacy in my composition. Among the advantages which I supposed a residence in the latter place would afford were the facilities offered of becoming acquainted with everything of importance which is going on in the Scientific World and by connecting myself with the University the possibility of some advantage of a pecuniary nature. To confess the truth I am mercenary enough to attach weight to the latter consideration though perhaps not sufficiently so to make a very great sacrifice of time in order to obtain it. Should what I have just advanced at all alter your view of the case I have no doubt but you will be kind enough to inform me of it. Before receiving your letter I had intended to ascertain if possible which college would be most suitable for a person of my age and *imperfect Classical Attainments* but having already experienced the advantage of being guided by your opinion I shall not on light grounds act in opposition thereto.
>
> I remain with the Greatest Respect

Yours very sincerely
Geo Green

This letter, written to Bromhead after a relationship of three years' standing, provides a welcome glimpse into Green's mind at an important moment in his life when he frankly discusses his attitudes and motives. It would appear that the rent of the family house would subsidize his college expenses, and that he had no intention of setting up Jane and her four children[27] in a family home to which he could return in the vacations. It has been suggested that they were in fact the 'temptations to mental dissipation' against which Bromhead warned Green,[28] but it is a moot question whether he knew of their existence. Regarding the possibility of a College Fellowship, for which celibacy was required, they were an embarrassment to Green, though not an insuperable barrier provided no marriage ceremony had taken place. Such common law relationships, both inside and outside the universities, were not unknown, and Bromhead was enough of a man of the world to accept such a situation had Green confessed to it.

The Bromhead correspondence is useful for the few indications it gives us of Green's mathematical thought, but from the personal point of view he remains largely an enigmatic figure. There is a clear change of style in the dozen letters written between 1830 and 1833. In contrast to his first letter, such later phrases as 'If you could spare an hour on any following day . . .' or 'Be so kind as to inform me . . .' are quite striking. A certain note of familiarity even creeps in when he refers to 'his apparent haste to get well quit of the two brats [i.e. the two memoirs] before the midsummer vacation . . .'[29] and an ingenuous echo of the Latin he was possibly now wrestling with, in his use of the Latin phrase 'rudis indigestaque molis' in his letter of 1833. These, however, are details. Green's basic approach to Bromhead remains one of respect and gratitude: 'I cannot conclude without expressing my gratitude for your kind assistance . . .'[30] 'The kindness which you have always manifested in giving me advice . . . and assistance . . .'[31] are examples of the frequent tributes to be found in the letters.

Green now prepared himself to go to Caius College. He would have to satisfy the examiners of his fitness to pursue an academic course, and in subjects other than mathematics. At this period the requirements were to possess a knowledge of the classics and the

Bible, and to be a baptized member of the Church of England, since only Anglicans could graduate and take their degree. William Tomlin stated that Green had relinquished his schooling in view of his mathematical superiority over his teachers: 'in consequence his literary requirements were not properly promoted: in this respect, he had when contemplating the probability of going to the University to pay some attention in his more mature years'. The examination, in March 1835, was over a year and a half away. This would give Green time for private study and coaching – perhaps from friends in Nottingham, so that his ignorance was not exposed in Cambridge. He also had to settle his financial affairs. His income would now depend on the rents from the mill and family house, the surrounding land and Gardens, the seven houses in Charles Street, and other real estate in Nottingham inherited from his parents. This would presumably guarantee him an income sufficient to pay his expenses in Cambridge and maintain his family in Sneinton.

George Green relinquished one more commitment before taking one of the two daily coaches to Cambridge. He had had the satisfaction of presenting a copy of the Essay, and now a copy of his first memoir, to the Bromley House Library, printed in Cambridge, as recorded in the Minute for 1 May 1833: 'The Thanks of the committee were voted to Mr. Green for his presentation of a Mathematical Investigation concerning the Laws of the Equilibrium of Fluids'. Finally, there is an entry for 5 August 1833: 'Mr. George Green's share transferred pro tempore to Mr. D. Shaw'. Membership of the Library was expensive, and Green probably felt that he had to husband his resources in view of uncertain expenses to be incurred at Cambridge. He presumably anticipated that his removal to Cambridge would be a lengthy – if not a permanent – one.

Chapter 7

An Undergraduate at Cambridge

George Green, writing to Bromhead in April 1833,[1] had referred to his 'inclination to go to Cambridge', and in listing his disadvantages he felt that he had 'seen too many winters'. When he arrived there in October of the same year, he had in fact seen forty. He went to Gonville and Caius College on the recommendation of Bromhead, who gave him letters of introduction to – in his own words – 'some of the most distinguished characters of the University'.[2] He entered first as a pensioner, becoming a scholar from 1834 to 1836, and took up residence in Gonville Court, where he stayed though he did not always occupy the same rooms, throughout his time at Cambridge.

Gonville and Caius College owes its double name to its double foundation. The first was in 1348, when Edmund Gonville of Norfolk obtained Letters Patent from Edward III for the foundation of 'a college of twenty scholars in the University of Cambridge'.[3] In 1557 John Caius (or Keys), eminent physician and humanist, was granted Letters Patent in the names of the Queen, Mary Tudor, and her husband, Philip of Spain, to refound the College. When within a year the Master of the College died, Caius was appointed in his place and, in due course, he revised the College statutes. Two are of particular interest. One relates to chapel attendance: in Caius's time it was 5 a.m. daily, though by the early nineteenth century this had become 7 a.m. A second statute required celibacy: 'We ordain that all members of your College, Master, fellows, scholars and pensioners be celibate and living in celibacy perpetual and honest for as long as they remain in College.'[4] Possibly through the professional eminence of its second founder, Caius College established – and still holds – a reputation for medical science. William Harvey, discoverer of the circulation of the blood, was one of its outstanding alumni.

John Caius extended the College buildings and site, and built the three famous Gates. As a 'Renaissance man', who had travelled and studied in Italy, he was imbued with a sense of the ancient classical virtues 'filtering through an Augustinian vision of humility'.[5] Thus the student coming to Caius College entered through the Gate of Humility. With study and perseverance he would pass through the Gate of Virtue; his efforts finally crowned, he would pass through the Gate of Honour, to receive his degree. In Caius's time the degree ceremony was presumably held in Great St Mary's, since the present Senate House, sited next to Caius College, was not built until 1773. A print of Caius Court in 1841 (the year after Green left Cambridge) shows the open gate, its pinnacle surmounted by a thriving mass of ivy – the usual concomitant of nineteenth-century historic buildings – with the roof of the Senate House and the towers of King's College beyond.

The entry for Green in the *Biographical History of Gonville and Caius College* reads as follows:

> *Green,[6] George: admitted pensioner. Oct. 6, 1833. (Son of George and Sarah Green, of Nottingham. Born July 14, 1793.)
>
> B.A. 1838 (4th wrangler). Scholar, L. Day 1834 to Michs 1836. Fellow, Michs 1839 to his death. A very distinguished mathematical physicist. In his youth he assisted his father in his business; first as a baker at Nottingham and afterwards as a miller at Sneinton. His father died in 1829; and, his mother having died before, he disposed of the business and prepared to enter college. He did not long enjoy his fellowship, as he went home in ill-health, in 1840, and died May 31, 1841. He was buried, with his parents, at Sneinton. His mathematical work was almost all done before graduation. His famous 'Essay on the application of mathematical analysis to the Theories of Electricity and Magnetism' was published at Nottingham by subscription, in 1828; but met with little success. It was largely owing to the help and encouragement of Sir E.T.F. Bromhead, at a period of much despondency, that he continued his studies and prepared to enter college. His mathematical works were edited by Dr. Ferrers in 1871.

The *Biographical History* lists thirty-three students who entered as pensioners during the year January–December 1833; there was no fixed entry date. Apart from Green, at forty, there were only one or two other older men, including an ex-naval doctor and Member of the Royal College of Surgeons. Otherwise all entrants were between seventeen and twenty-three, and of the thirty-three entrants, five did not graduate. Ten came from public schools and four had been privately educated. Fathers' standing included surgeons, barristers, 'gentlemen' and an army colonel. The majority of students appear to have come from reasonably prosperous middle-class families, and to have been educated at grammar or private schools.

Green moved four times during his six years in Cambridge, occupying one of the two ground- or first-floor rooms on a staircase in Gonville Court. These were some of the best rooms in the College: the ground floor rooms now include those allocated to Fellows of the College. As a pensioner Green would have paid full fees, until his promotion to scholar from Lady Day 1834. Lady Day actually falls on 25 March; it signified the end of the first half of the academic year, which started the previous Michaelmas, 29 September. Lady Day also marked the end of the first half of the financial year, and College scholarships would normally commence on this day. The second half-year ran from Lady Day, the Feast of the Annunciation of the Virgin Mary and the patronal festival of Gonville's foundation, through the May term and the long vacation to the following Michaelmas.

Green presumably had the means to provide himself with comfortable accommodation and was not confined to an attic room, as was John Venn some twenty years later; during one cold winter, the water froze in his jug every night.[7] No water was laid on in the College, but there were two or three outside pumps. Cold baths could be taken – some colleges provided a bathhouse – otherwise students could take dips in the Cam before Chapel at 7 a.m. When, on a dark, wet December evening, a modern visitor walks through Gonville Court, now lit by electric lamps, and wonders how the courts were lit in Green's time, a hundred and sixty years ago, it is intriguing to find a clue in the Minute Book or 'Gesta', which recorded the decisions of the Master and Senior Fellows. On 31 October 1838 the Fellows resolved 'to have a gas lamp placed at the door of the Porter's lodge instead of the oil lamp as at present'.

Chapel and Hall were the meeting points for the College community – pensioners, scholars and Fellows alike. Dinner, the one meal taken in Hall, was at 4 or 5 p.m. On 13 May 1837 the 'Gesta' records the decision 'that the allowance of meat for the undergraduates and bachelors during the ensuing Michaelmas quarter be a pound and a half a day for each'. Venn records how joints of meat were passed round the table and 'hacked' by each student, serving himself.[8] Soup, fish and game were unknown; sweets and cheese had to be ordered specially. The College servants were paid by 'beer orders' redeemable at the buttery. Venn was alarmed at the number of these, and asked one individual how he managed to cope. 'We works it off, sir,' was the reply, 'we works it off.'[9]

Members of the College were distinguished from the townspeople by the wearing of cap and gown, as shown in numerous contemporary engravings. Lord Byron, as an undergraduate, was entitled to wear a black silk top hat instead of a mortarboard, since he was of aristocratic status. Edward and Charles Bromhead, as sons of a baronet, were likewise so entitled.

Green's first examination was the College Scholarship examination held in the first year. This, like the 'Previous Examination' or 'Little-Go' held in year two, was in two parts and comprised three areas of study. According to Venn, the Classical and Theological Examination was held in the first year at the end of the Lent term and Mathematics at the end of the first May term. As Green wrote to Bromhead on 22 May 1834:[10] 'We are now in the midst of an examination here and of course have our time almost completely engrossed.' Fortunately for Green, the mathematics would not present a problem: in fact he gained the first year mathematical prize. On his own confession he possessed, as previously stated, 'little Latin, less Greek'[11] and had enquired of Bromhead 'which College would be most suitable for a person of my age and *imperfect Classical Attainments*',[12] but he may have overestimated the effort needed to attain a successful result in the first half of the examination, since although the scholar's status was awarded on the joint result of both examinations, the standard was probably not very high. A similar procedure took place in the second year for the Previous Examination. This was a necessary step for the degree of BA, which was awarded to the average run of student after a suitable period of residence. But even the brilliant young William Thomson, writing home from St Peter's College in 1843, was not taking any chances: 'This evening I have been working at Paley and Xenophon, keeping steadily before my mind the fear of being plucked.'[13] At nineteen, however, Thomson, though still mindful of the importance of the 'Little-Go', had been translating and parsing Greek at the age of twelve; Green, at twelve, had been working in his father's bakery, and he was now a middle-aged man.

George Green would get little help for this study. Venn states that two hours of lectures were held every morning during the first two years leading up to the Previous Examination. There was no entrance examination to the College, and the same lecture was given to all students: the classical scholar would be sitting next to one with the most elementary Greek; the mathematician next to one making a first acquaintance with Euclid. Indeed, Venn declared that the great advantage of the long vacations was that they were unhindered by lectures and one could get on with one's work. So students found their own coaches and paid for their tuition, or they studied on their own, or learnt from each other – on afternoon walks, in the coffee house, or in each other's rooms.

University professors in Cambridge, then as now, had no statutory obligation to lecture, and appointment to a Chair was still, on occasion, a symbol of status. Charles Babbage, for example, was appointed Lucasian Professor of Mathematics in 1828, but he never visited Cambridge to lecture. By the 1820s, however, professors were no longer expected to treat their appointments as sinecures. George Peacock was severely criticized for retaining the Lowndean Chair after his translation to Ely as Dean in 1839.[14]

Some of the outstanding figures, however, published works which became valuable university textbooks. Whewell published *An Elementary Treatise on Mechanics* in 1819 (a copy was acquired by Bromley House in 1828) and *A Treatise on Dynamics* four years later. In 1826 George Biddell Airy, Plumian Professor of Astronomy, published his *Elementary Tracts on Physical Astronomy, figure of the earth, Procession and Nutation, and Calculus of Variations*, which was widely used as a university textbook. Bromhead sent a copy to Green, which the latter acknowledged in his letter of 23 May 1832.[15] Peacock wrote a *Treatise on Algebra* in 1830, with a revised edition in 1842. As a past member of the Analytical Society, he had been largely responsible for the publication of the book of examples based on the Society's translation in 1816 of Lacroix's *Traité du calcul différentiel et du calcul intégral.*[16] College Fellows were under more pressure to lecture in addition to their tutoring, though even this may not have been onerous. Private tuition to students was a means of increasing income, and some tutors were excellent in an informal way. It is gratifying to record the comments of past students on George Peacock's teaching as Fellow:

> 'While his extensive knowledge and perspicuity as lecturer maintained the high reputation of his college and commanded the attention and admiration of his pupils, he succeeded to an extraordinary degree in winning their personal attachment by the uniform kindliness of his temper and disposition . . .'
>
> 'His inspection of his pupils was not minute, far less vexatious, but it was always effectual and at all critical points of their career, keen and searching. His insight into character was remarkable.'[17]

Green had arrived in Cambridge with letters of introduction from Bromhead. This was quite a normal procedure: William Thomson, for example, had a dozen such letters when he departed for Paris after taking his degree. Likely addressees would include Whewell and Peacock, and possibly Airy, and there would have been others, as Bromhead had a number of Cambridge contacts. Green's thoughts about whether and when to avail himself of them must have been mixed. He was forty years old, but he was arriving in Cambridge as an undergraduate – older than students and Fellows, and in fact the contemporary of these eminent Cambridge figures. His academic humility had revealed itself when he had refused Bromhead's invitation to accompany him to Cambridge earlier that year. As for those who constituted 'the Chivalry of British Science', Whewell now held the Chair of Mineralogy and in a few years would become Master of Trinity; Peacock would later be appointed Lowndean Professor of Astronomy; and Airy would soon depart to become Astronomer Royal at Greenwich Observatory.

These were daunting figures for Green, despite the fact that he would have the satisfaction, before the end of his first term, of having his memoir on the equilibrium of fluids published in the *Transactions of the Cambridge Philosophical Society* and knowing that his second memoir on ellipsoids had been read the previous May. Whewell and Peacock, as officers of the Society, would thus have known of his worth. Indeed, his reputation as a mathematician would grow in Cambridge. He would be singled out as the future Senior Wrangler of his year, and Harvey Goodwin would later declare his superiority over all others in the University.[18] The *Caius Biographical History*, however, lists him as 'son of a miller', and Green had worked as a miller himself up to three years before his arrival in Cambridge. This, in essence, should have been no handicap – the medieval colleges had their poor scholars, and in later years they had their sizars. Isaac Milner, son of a weaver, had been one. Robert Murphy – of whom more below – was the son of an impoverished Irish shoemaker, and even William Whewell was the son not of a gentleman but of a master carpenter in Lancaster.

Green now had the status of a gentleman, with the means to keep himself in reasonable comfort, but he had not had one significant advantage, shared by the others: he had not arrived at Cambridge at the age of seventeen or eighteen and spent his formative years in college.[19] The men of Cambridge, whatever their background and their internal divisions of status, constituted a freemasonry, and to the outsider they may well have appeared a closed and privileged society. Green had confessed earlier to Bromhead that he had a 'tolerable share of obstinacy in [my] disposition'.[20] Nottingham people have the reputation of having an independent turn of mind, and Green would have withdrawn with dignity from a ready, patronizing manner.

It is as well, perhaps, that Green was unaware of Bromhead's thoughts in this matter, which he describes in his letter of 1845:

> We also conversed about Cambridge and on another occasion when he inclined to go there. I recommended my own college of Gonville and Caius at which with good conduct and a high degree he might be certain of a fellowship – I also gave him letters of introduction to some of the most distinguished characters of the University that he might keep his object steadily in view under some awe of their names and look upwards, not of course with any view of trespassing on the social distinctions of our University, in my time much more marked than at present, but that he might venture to ask advice under any emergency.[21]

This inevitably sounds condescending, though Bromhead was in fact retailing circumstances of fifteen years before, and at a time when he may have been wishing to distance himself slightly from his past association with Green.

As far as is known, Green sent only one letter to Bromhead from Cambridge, and it appears to be the last in the Bromhead archive. This begs the question whether he continued his visits to Thurlby during the vacations, since letter-writing was now difficult under pressure of work. Yet the lack of known communication after 1834, and the general tone of Bromhead's letter of 1845, suggests that the two men may not have maintained their former association. However that may be, Green's letter to Bromhead of 22 May 1834[22] deals first, like his previous ones, with the publication of his memoirs, and it is written in what became his usual style – more fluent and relaxed than the long-winded and rather tortuous expression of the earlier letters. The first paragraph is concerned with the proofreading of the memoir on the vibrations of pendulums in fluid, which had been sent to Edinburgh. Green had been uncertain about procedure in returning it, and on Bromhead's advice he had consulted Professor Whewell. He states that he is not interested in having extra copies from Edinburgh, because 'the paper is a rather unimportant one', and he is more concerned over the fate of the previous memoir on ellipsoids. This had been read twelve months previously – in May 1833, before his arrival in Cambridge – but its publication was still uncertain:

> This last paper I fear will scarcely be printed on account of its author being an undergraduate I have had no direct communication to this effect from the Society but have heard . . . that this circumstance forms a most serious objection to its admission into the valuable Miscellany.

The memoir on ellipsoids was in fact published in the 1835 *Cambridge Transactions* – a rare tribute, and a confirmation of the respect that Green was attracting in Cambridge. Green then comments on his immediate plans. Examinations would end the following day, and he would leave Cambridge 'early on Wednesday morning next and think I shall spend the long Vacation in the neighbourhood of Nottingham. I am not yet certain where but anything directed for me at Mr. W.Tomlin's High Pavement Nottingham would be sure to find me.' This letter is dated 22 May. Four days previously,

1. Isaac Milner 1750–1820

A Treatise

UPON

ANALYTICAL MECHANICS;

BEING THE FIRST BOOK

OF THE

MECANIQUE CELESTE

OF

M. LE COMTE LAPLACE,

Peer of France, Member of the Institute, of the Bureau of Longitude, &c. &c.

TRANSLATED AND ELUCIDATED

WITH

Explanatory Notes,

BY THE REV. JOHN TOPLIS, B. D.

FELLOW OF QUEEN'S COLLEGE, CAMBRIDGE,

Nottingham;

PRINTED BY H. BARNETT.

SOLD BY LONGMAN, HURST, REES, ORME, AND BROWN; AND CRADDOCK AND JOY, LONDON; AND J. DEIGHTON AND SONS, CAMBRIDGE.

1814.

2. The title page of Toplis's translation of Laplace

AN ESSAY

ON THE

APPLICATION

OF

MATHEMATICAL ANALYSIS TO THE THEORIES OF ELECTRICITY AND MAGNETISM.

BY

GEORGE GREEN.

Nottingham:

PRINTED FOR THE AUTHOR, BY T. WHEELHOUSE.

SOLD BY HAMILTON, ADAMS & Co. 33, PATERNOSTER ROW; LONGMAN & Co.; AND W. JOY, LONDON; J. DEIGHTON, CAMBRIDGE;

AND S. BENNETT, H. BARNETT, AND W. DEARDEN, NOTTINGHAM.

1828.

conducting bodies, and is an immediate consequence of what has preceded. For let x, y, z, be the rectangular co-ordinates of any particle p in the interior of one of the bodies; then will $-\left(\frac{dV}{dx}\right)$ be the force with which p is impelled in the direction of the co-ordinate x, and tending to increase it. In the same way $-\frac{dV}{dy}$ and $-\frac{dV}{dz}$ will be the forces in y and z, and since the fluid is in equilibrium all these forces are equal to *zero*: hence

$$0=\frac{dV}{dx}dx+\frac{dV}{dy}dy+\frac{dV}{dz}dz=dV,$$

which equation being integrated gives

$$V=\text{const.}$$

This value of V being substituted in the equation (1) of the preceding number gives

$$\rho=0,$$

and consequently shows, that the density of the electricity at any point in the interior of any body in the system is equal to *zero*.

The same equation (1) will give the value of ρ the density of the electricity in the interior of any of the bodies, when there are not perfect conductors, provided we can ascertain the value of the potential function V in their interior.

(3.) Before proceeding to make known some relations which exist between the density of the electric fluid at the surfaces of bodies, and the corresponding values of the potential functions within and without those surfaces, the electric fluid being confined to them alone, we shall in the first place, lay down a general theorem which will afterwards be very useful to us. This theorem may be thus enunciated:

Let U and V be two continuous functions of the rectangular co-ordinates x, y, z, whose differential co-efficients do not become infinite at any point within a solid body of any form whatever; then will

$$\int dxdydz\, U\delta V+\int d\sigma\, U\left(\frac{dV}{dw}\right)=\int dxdydz\, V\delta U+\int d\sigma\, V\left(\frac{dU}{dw}\right);$$

the triple integrals extending over the whole interior of the body, and those relative to $d\sigma$, over its surface, of which $d\sigma$ represents an element: dw being an infinitely small line perpendicular to the surface, and measured from this surface towards the interior of the body.

To prove this let us consider the triple integral

$$\int dxdydz\left\{\left(\frac{dV}{dx}\right)\left(\frac{dU}{dx}\right)+\left(\frac{dV}{dy}\right)\left(\frac{dU}{dy}\right)+\left(\frac{dV}{dz}\right)\left(\frac{dU}{dz}\right)\right\}.$$

The method of integration by parts, reduces this to

$$\int dydz\, V''\frac{dU''}{dx}-\int dydz\, V'\frac{dU'}{dx}+\int dxdz\, V''\frac{dU''}{dy}-\int dxdz\, V'\frac{dU'}{dy}$$
$$+\int dxdy\, V''\frac{dU''}{dz}-\int dxdy\, V'\frac{dU'}{dz}-\int dxdydz\, V\left\{\frac{d^2U}{dx^2}+\frac{d^2U}{dy^2}+\frac{d^2U}{dz^2}\right\};$$

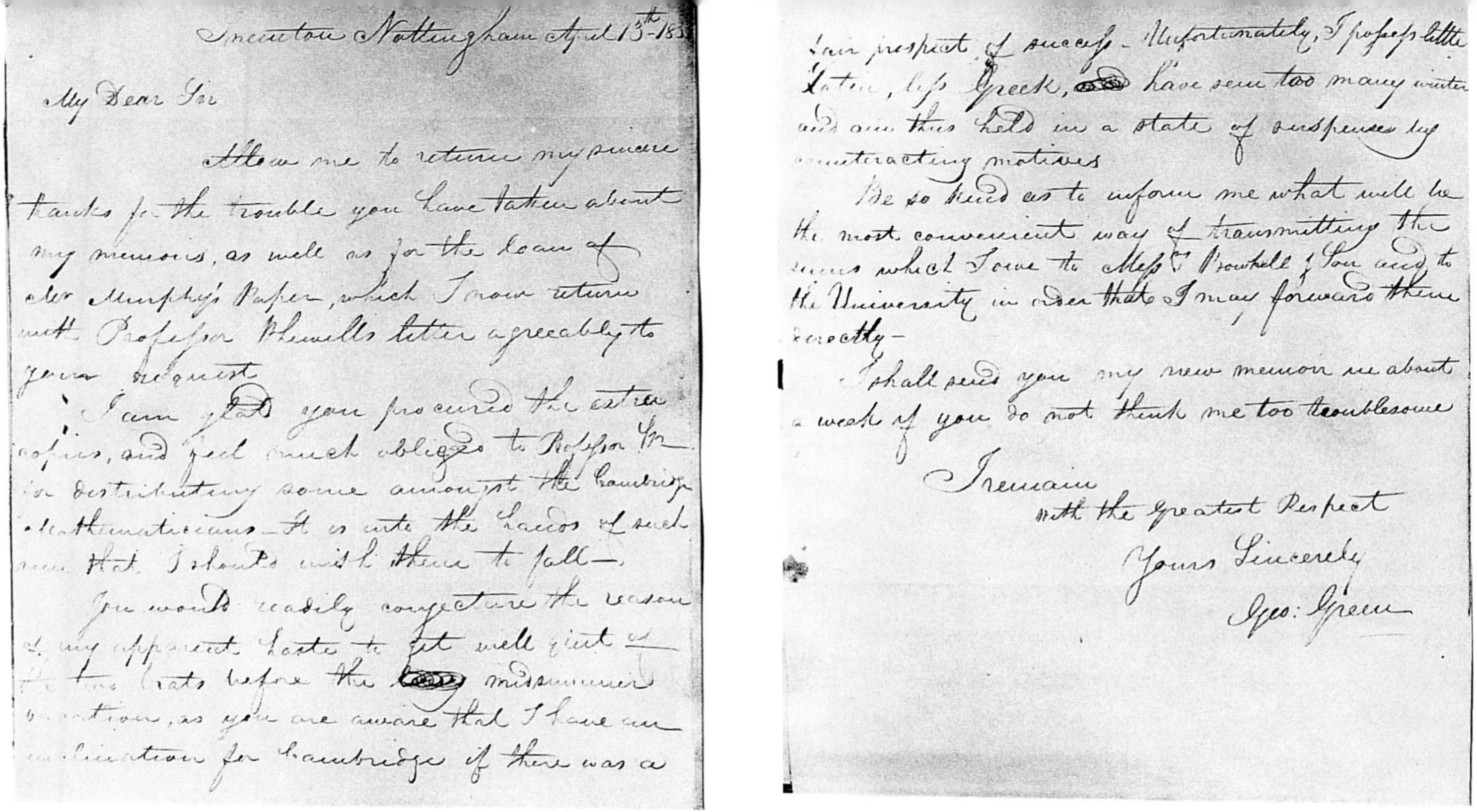

Sneinton Nottingham April 13th 1833

My Dear Sir

Allow me to return my sincere thanks for the trouble you have taken about my memoirs, as well as for the loan of Mr Murphy's Paper, which I now return with Professor Whewell's letter agreeably to your request.

I am glad you procured the extra copies, and feel much obliged to Professor [illegible] in distributing some amongst the Cambridge Mathematicians. It is into the hands of such men that I should wish them to fall.

You would readily conjecture the reason of my apparent haste to get well quit of the two tracts before the midsummer vacation, as you are aware that I have an inclination for Cambridge if there was a fair prospect of success. Unfortunately, I profess little Latin, less Greek, have seen too many winters and am thus held in a state of suspense by counteracting motives.

Be so kind as to inform me what will be the most convenient way of transmitting the sums which I owe to Messrs Bowtell & Son and to the University in order that I may forward them correctly.

I shall send you my new memoir in about a week if you do not think me too troublesome.

I remain
with the Greatest Respect
Yours Sincerely
Geo: Green

5. One of Green's letters to Bromhead, dated April 13th 1833

6. Bromley House c. 1880 (third from left)

7. Gate of Honour, Caius College Cambridge, 1841

8. Edward Bromhead 1789–1856

9. Charles Babbage 1792–1871

. John Frederick Herschel 1792–1871

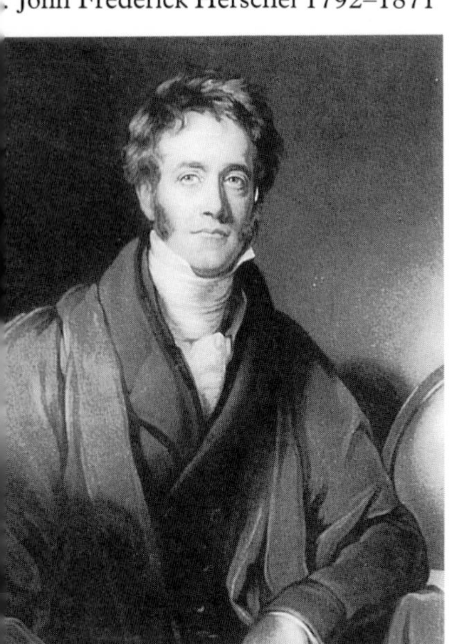

11. George Peacock 1791–1858

12. William Whewell 1794–1866

13. Robert Murphy 1806–1843

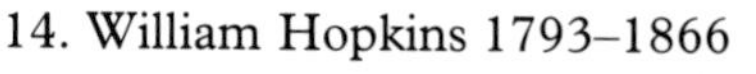

14. William Hopkins 1793–1866

15. William Thomson (Lord Kelv
1824–19

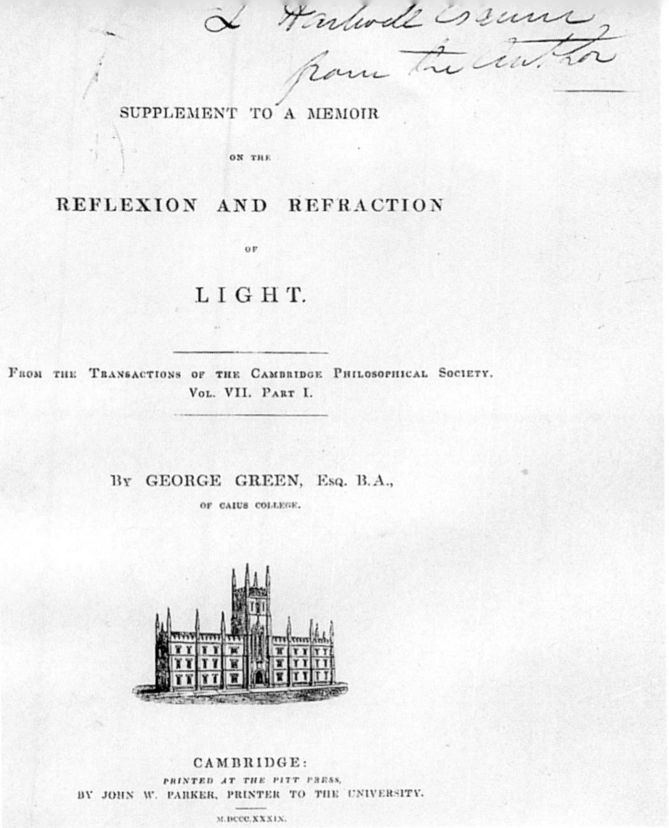

SUPPLEMENT TO A MEMOIR

ON THE

REFLEXION AND REFRACTION

OF

LIGHT.

FROM THE TRANSACTIONS OF THE CAMBRIDGE PHILOSOPHICAL SOCIETY.
VOL. VII. PART I.

BY GEORGE GREEN, ESQ. B.A.,
OF CAIUS COLLEGE.

CAMBRIDGE:
PRINTED AT THE PITT PRESS,
BY JOHN W. PARKER, PRINTER TO THE UNIVERSITY.
M.DCCC.XXXIX.

16. Title page of Green's second memoir on light, 1839

17. Mrs Jane Moth, George Green's eldest daughter

18. The British Association's tribute to Green, 1937

19. Green's Mill before restoration, 1975

20. The Mill restored, 1985

21: The Dean of Westminster dedicating the Plaque to George Green 16 July 1993 in front of statue to Isaac Newton.

22: Left to right: Professor Freeman Dyson, Professor Lawrie Challis, and Professor Julian

Jane Smith had given birth to her fifth child, Catherine. The 'neighbourhood of Nottingham' may be a reference to Sneinton, not then a part of the town; and giving Tomlin's address as 'poste restante' preserves his family's anonymity, were he to be staying there during the vacations. Then comes the final paragraph. Green may have felt some reserve in meeting some of the 'distinguished characters' amongst Bromhead's acquaintance, but in his general college life he appears to have had none:

> I am very happy here and I fear too much pleased with Cambridge. This takes me in some measure from those pursuits which ought to be my proper business but I hope on my return to lay aside my freshnesses and become a regular steady Second Year Man.

Cambridge was not without its problems. Town and gown were sometimes at odds, but Green was 'a reading man',[23] and not of the student population concerned in these youthful disputes. The College community at Caius – about a hundred and twenty students and some twenty Fellows – obviously provided a society which Green found congenial. In certain circumstances there could be strict formality between tutors and students, but Green – an older man, and one with increasing mathematical stature – was in a position to associate with the Fellows, either in his own rooms or on occasion in the Fellows' Combination Room. Not least, the quiet and graciousness of the buildings, and the pleasant and airy situation of the colleges backing on to the meadows and the river, would have presented an agreeable contrast to Nottingham, with its busy, congested marketplace, its slum areas and its frequent riots – not to mention the possibly distracting domestic responsibilities which were renewed during the vacations.

By the end of his second year in Cambridge, Green had taken the Previous Examination and finished his Greek and Latin studies. Now, in his third year, he could devote himself entirely to mathematics and prepare himself in the four terms ahead for the ordeal of the Senate House examination.

George Peacock, writing in 1841, divided students into two classes:[24] those with wealth and position, for whom success was not important, and those whose circumstances compelled them to graduate before seeking a livelihood in one of the professions: the law, medicine or the Church. For the one category, a year or two at Cambridge was the equivalent of the eighteenth-century Grand Tour. For the other, a degree of any value, leading to academic honours and a career, meant entry into the mathematical 'honours' school and sitting the Mathematical Tripos, held in the Senate House in the January of a student's fourth year. College competition for the honour of producing the Senior Wrangler, or top man of the year, rivalled that of being head of river in the annual 'bumps' races. Names of the successful candidates were published in three classes: the Wranglers (first-class honours), the Senior Optime (second-class students) and the Junior Optime (third class students). In each class the names were published in order of merit. The Senior Wrangler was often spotted quite early. A fellow freshman with Thomson recorded that after only five days' residence it was currently reported in the College that 'Thomson would be Senior Wrangler'.[25] Thomson himself refers to meeting Arthur Cayley, 'who is to be Senior Wrangler this year'[26] – and so he was in 1841.

The Senate House examination was indeed a gruelling experience. Two examiners – themselves young Fellows in their early twenties – set, invigilated and marked the papers for a five-day examination. The papers on the first three days dealt with the more general groundwork of the course; those on the last two days presented problems which would identify the more able students.[27] Whewell wrote to his sister in 1820,[28] describing the fatigue of invigilating for five consecutive days in the first week of January from 7 a.m. to 5 p.m. in an unheated Senate House – its high ceilings and interior decoration precluding any interior heating arrangements – and then checking papers every night. Final results were published within ten days.

As for the candidates, they too suffered the same conditions, writing as fast as they could in order to answer as many questions as possible. Up to the year 1800, questions had been dictated singly by the moderator. As soon as one of the candidates had solved the problem, he raised his hand, all work stopped and the next question was dictated. From 1801 onwards, however, the examination papers were printed.[29] The top place in the examination was virtually won by the candidate who could get most down on paper. This system 'was not adapted either to cultivate originality or to advance mathematical science'.[30] On the contrary, it favoured the candidate with a retentive memory who had the good fortune, and the finance, to have been tutored by one of the leading coaches in the University.

Coaches – such as Hopkins of St Peter's, or Hymers of St John's – played a major role in teaching the more affluent students. William Hopkins, in the course of some twenty years, was credited with producing over two hundred Wranglers, of whom seventeen were Senior Wranglers.[31] Thus a second accolade for the brilliant student was to have been a pupil of Hopkins. William Thomson was one such, as were Stokes, Blackburn, Tait and Clerk Maxwell. Hymers of St John's was the arch-'crammer', who 'could tell to a nicety the kind of work that would pay in the Senate House Examination'.[32] When William Thomson of St Peter's was beaten into second place by Parkinson of St John's, Whewell wrote of him: 'Thomson of Glasgow is much the greater mathematical genius: the Senior Wrangler was better drilled.'[33]

The supremacy of St John's in the Senate House examination was outstandingly demonstrated in 1837, when Green took the Tripos. In that year all three top places were taken by Johnians – Griffin, Sylvester and Brumell, of whom only Sylvester achieved any fame as a mathematician. And they were all, of course, young men. This result must have caused considerable surprise in Cambridge, since Green, like Cayley and Thomson later, had been singled out as the Senior Wrangler of his year. Green's result was obviously disappointing – not least, in all probability, to himself – and various people sought to find a reason: in H.G. Green's case, with some justification. 'Green and Sylvester were the first men of the year,' wrote Harvey Goodwin, 'but Green's want of familiarity with schoolboy mathematics prevented him from coming first in a time race. It was a surprise to everyone to find Griffin and Brumell had beaten him.'[34] N.M. Ferrers wrote of this result:

> It is hardly necessary to say that this position, distinguished as it was, most inadequately represented his mathematical power. He laboured under the double disadvantage of advanced age and of inability to submit entirely to the course of systematic training needed for the highest place in the Tripos.'[35]

Green had also laboured under a third disadvantage which dogged him until he had taken the Previous Examination: the acquisition of Latin and Greek almost from scratch.

Tomlin had his own explanation for Green's result:

> In the regular course he tried for University honours where to the disappointment of his friends he was only fourth wrangler, which circumstance was understood by them to arise from the easiness of the subjects given to the competitors and which is proved by the number of wranglers of that year being the greatest ever known.[36]

To Green's friends and contemporaries in Nottingham, such a facile explanation of his result may have been convincing.

Having spent his Christmas vacation in college preparing for his final examinations, Green departed on 27 January. The date of his return is not recorded, though he left college as usual at the end of the Michaelmas term. He would have needed to relax after the strain of the Tripos examination. It had considerable significance for him – on it depended his hope of a Fellowship, for which he had forgone the stability and comforts of family life and for which, perhaps, he denied his children legitimacy. The disappointment of coming fourth in the final result must have been considerable. He cannot have been unaware of the speculation surrounding him, of his reputation in the University, or, perhaps, of the disappointment of his friends in Nottingham.

From Jane Smith he probably got affection and support, though for her the circumstances cannot have been easy. Nine days after Green's return to Nottingham at the end of January, she gave birth to her sixth child. She was baptized Elizabeth Green and, like her brothers and sisters, registered as illegitimate. The baptism took place at St Stephen's, Sneinton, and Jane's address was given as Windmill Hill. But as it was term time, Green would be obliged to keep residence and return to Caius. Perhaps he was quite relieved to do so. A household of six children, whose ages ranged from a few weeks to thirteen, could scarcely have been restful.[37] In contrast, Cambridge offered its quiet courts and peaceful river, its leisurely pace and cultured living. Above all, it now presented Green with limitless time – and no commitments – to continue his studies.

Chapter 8

A Fellowship at Caius College

George Green, now a graduate of the University, was elected a Fellow of the Cambridge Philosophical Society on 6 November 1837. This was not the first time George Green's name had been mentioned in the General Meeting Proceedings, since they had recorded the receipt of the Essay and three subsequent investigations. Green could therefore be assured of a welcome into the Society, and a reception infused with respect for his academic ability from Whewell, Peacock and Hopkins, all of whom were members of the Council. Green, at forty-four – and for the first time in his life – now had ample time to read, to think and to write. Whereas the next two years may have been frustrating personally, as he waited to be considered for a Caius Fellowship at the monthly meetings of the College Senior Fellowship, they were a time of considerable intellectual activity. The titles of his first four investigations – three on electricity and one on hydrodynamics – are known. It might be useful here to give those of the Cambridge period. The sequence of the papers is as follows:

'On the Motion of Waves in a Variable Canal of Small Width and Depth', read 15 May 1837, printed in the *Transactions* 1838. Quarto, 6 pages.

'On the Reflexion and Refraction of Sound', read 11 December 1837, printed in the *Transactions* 1838. Quarto, 11 pages.

'On the Laws of the Reflexion and Refraction of Light at the common Surface of two non-crystallized Media', read 11 December 1837 printed in the *Transactions* 1838. Quarto, 24 pages.

'Note on the Motion of Waves in Canals', read 18 February 1839, printed in the *Transactions* 1839. Quarto, 9 pages.
'Supplement to a Memoir on the Reflexion and Refraction of Light', read 6 May 1839, printed in the *Transactions* 1839. Quarto, 8 pages.
'On the Propagation of Light in Crystallized Media', read 20 May 1839, printed in the *Transactions* 1839. Quarto, 20 pages.

Green's move to Cambridge was a turning point in his life – as is revealed, not least, in the subject matter of these memoirs. There had been the three Nottingham papers on electricity, but this was not a popular subject of study in Cambridge.[1] Before his arrival there, however, Green had already followed Whewell's advice to devote attention to other topics, the first of which was hydrodynamics, on which he finally wrote three papers.The Edinburgh paper deals with the problem of 'the motion of an inelastic fluid agitated by the small vibrations of a solid ellipsoid, moving parallel to itself'. This problem is put in its context in the opening paragraph:

> PROBABLY no department of Analytical Mechanics presents greater difficulties than that which treats of the motion of fluids; and hitherto the success of mathematicians therein has been comparatively limited. In the theory of waves, as presented by MM. Poisson and Cauchy, and in that of sound, their success appears to have been more complete than elsewhere; and if to these investigations we join the researches of Laplace concerning the tides, we shall have the principal important applications hitherto made of the general equations upon which the determination of this kind of motion depends. The same equations will serve to resolve completely a particular case of the motion of fluids, which is capable of a useful practical application; and as I am not aware that it has yet been noticed, I shall endeavour, in the following paper, to consider it as briefly as possible.[2]

In the first Cambridge paper of 1838, Green applies the 'powers of Analysis' to a particular case, susceptible of solution by this means in an otherwise difficult field, 'that of an indefinitely extended canal of small width and depth, both of which may vary very slowly, but in other respects quite arbitrarily'. In the second Cambridge paper of 1839, he links the suppositions of the earlier paper to one contained in the Seventh Report of the British Association for the Advancement of Science (1836) – the report of the Association's Committee on Waves. The investigations into the phenomenon of waves and tides attracted much interest. William Whewell, amid his multifarious interests, had headed a similar committee set up by the Royal Society, and published a series of papers in its *Transactions*. The British Association Report was presented by Sir John Robison, Secretary of the Royal Society of Edinburgh, and John Scott Russell, Fellow of the Royal Society of Edinburgh. Green re-examines his earlier suppositions in the light of their report.

> These suppositions are not always satisfied in the propagation of the tidal wave, but in any other cases of propagation of what Mr. Russel [*sic*] denominates the 'Great Primary Wave', they are so, and his results will be found to agree with our theoretical deductions.[3]

Meanwhile, Green was involved with problems in sound and light, optics in particular being a widely discussed subject at this time. Green first produced two memoirs on sound and light, which were read on the same date and published in 1838. N.M. Ferrers, having recommended that the two papers should be studied together, proceeds to give a brief analysis of their scope:

> The question discussed in the first is, in fact, that of the propagation of normal vibrations through a fluid. Particular attention should be paid to the mode in which, from the differential equations of motion, is deduced an explanation of a phenomenon analogous to that known in Optics as Total internal reflection when the angle of incidence exceeds the critical angle. . . . The immediate object of the next paper, 'On the Reflexion and Refraction of Light at the common surface of two non-crystalline [*sic*] media', is to do for the theory of light what had been done for the theory of sound. . . . But this paper has an interest extending far beyond this subject. For the purpose of explaining the propagation of transverse vibrations through the luminiferous ether, it becomes necessary to investigate the equations of motion of an elastic solid. It is here that Green enunciates the principle of the Conservation of Work, which he bases on the assumption of the impossibility of perpetual

> motion. This principle he enunciates in the following words: 'In whatever manner the elements of any material system may act upon each other, if all the internal forces be multiplied by the elements of their respective directions, the total sum for any assigned portion of the mass will always be the exact differential of some function'. This function . . . is what is now known under the name of Potential Energy.[4]

Green's work in this area was contemporary with that of Fresnel and Cauchy in France,[5] and that of MacCullagh in Ireland. The cross-fertilization of ideas, as traced by Whittaker, indicates that Green's work, through publication in the *Transactions*, was now in circulation.

Green published a Supplement to this memoir in 1839. His major work on the subject of light, however, was his last, 'On the Propagation of Light in Crystallized Media', which, after the Essay, is probably his most important paper, and in which:

> the principle of Conservation of Work is again assumed as a starting point and applied to a medium of any description. Fresnel's supposition, that the vibrations affecting the eye are accurately in front of the wave, is then introduced, and a complete explanation of the phenomena of polarization is shown to follow.[6]

The influence of Cambridge on Green is evident in another aspect of the memoirs, that of style. A number of writers have commented on this, considering his general characteristics to be conciseness and elegance.

> Green is pellucid, with writing aimed directly at the point and not obscurely wrapped round it: one must put this down to his French reading. Green's notation is clear and consistent, not always chopping and changing as in Murphy and Lagrange.[7]

H.G. Green finds in Green's style direct evidence of one source for his Essay in the 'conciseness of style which strongly resembles that of Toplis in his notes'[8] – Toplis himself being immersed, through his translations, in the texts of Laplace and his contemporaries. Whittaker, writing in 1910, is a third writer to recognize the influence of the French analysts on Green's style:

> Green undoubtedly received his own early impressions from this

> source [i.e. 'the works of the great French analysts'], chiefly Poisson; but in clearness of physical insight and conciseness of exposition he far excelled his masters: and the slight volume of his collected papers has to this day a charm which is wanting in their voluminous writings.[9]

J.J. Cross, in the first quotation, notes the clarity and consistency of Green's notation. He also quotes a tribute from George Gabriel Stokes. 'Mr. Green's memoirs are very remarkable, both for the elegance and rigour of the analysis, and for the ease with which he arrives at most important results.'[10] A close reading of the text of the Essay and the memoirs, however, reveals some change of style during their eleven years of composition. In the Nottingham works, the sentences are long, the style is complex, the vocabulary is polysyllabic; concentration is needed to hold the thread through the long paragraphs. One recalls Green's first letters to Bromhead, which are in strong contrast to the later letters, with their greater ease and conciseness of style. So it is with the memoirs.

The two Nottingham papers published in Cambridge were 63 and 35 pages long respectively, and the first had already been pruned considerably before publication. The Cambridge papers are much shorter. Textual matter is much reduced, the mathematical reasoning emerges clearly, and the whole radiates the intellectual delight and 'charm' noted by Whittaker. The years of association with students and College Fellows, and in particular with the leading figures of the University in the Cambridge Philosophical Society, undoubtedly affected Green's writing. Increasing confidence led to the shedding of unnecessary justification or explanation in the text. The cut and thrust of argument with keen minds – with Hopkins or Whewell outside Caius, Murphy, Paget or O'Brien within – sharpened his style, and statements and conclusions are enunciated with greater authority.

In writing these later investigations, Green had none of the earlier problems he had experienced in Nottingham. Cambridge University in the 1830s was emerging from its isolation; in terms of mathematical analysis, the battle waged by Toplis and then by the Analytical Society had been won a decade before, and Deightons were now advertising books by the French analysts in their catalogues.[11] Green could purchase these easily; he also had access to libraries. Particularly valuable were the resources of the Cambridge Philosophical Society, which regularly received *Bulletins* and *Transactions* from organizations at home and abroad.

Since its inception in 1820 the Society had sent its *Transactions*, generally on a reciprocal basis, to the Royal Societies of London and Edinburgh, the Irish Academy, the Astronomical, Geological and Linnean Societies and others, as well as to the Institut de France and the Département de la Marine in Paris, and to universities and academies in Brussels, Geneva, Lisbon, Amsterdam, Leyden, Munich, Göttingen, Königsberg, Berlin, Vienna, Uppsala, St Petersburg, Boston and Washington, to name the most prestigious. As a result, the Society's Library soon rivalled the other Cambridge libraries in its scientific content, and it now forms a significant element in the Periodicals Section of the main University Library. Copies of the *Transactions* were also sent as 'Presents' to individuals abroad – to Bessel and Schumacher, to Poggendorf in Berlin (his *Annalen* were received by the Society), and to Jacobi in Königsberg. Biot and Ampère had been made honorary members of the Society on its inauguration in 1820, and presumably received copies of its publications; in the next decade the Council Minutes show that Jacobi, Poisson, Cauchy and Arago also received copies – Whewell undertook to send Arago's copy, since he appears to have acted as foreign correspondent to the Society. Green therefore had the opportunity to keep abreast of the most recent publications – and, when appropriate, to communicate with individuals.

There was a certain amount of individual correspondence. It would appear that towards the end of his time in Cambridge, Green was in touch with at least one fellow investigator abroad. In 1976 Sotheby's sold at auction copies of six of Green's memoirs,[12] each inscribed: 'Professor Jacobi, from the Author', in Green's unmistakable hand.[13] It may be assumed that Green, with his reserve and quietness of manner, would have sent his papers to someone of Jacobi's reputation only on request, or at the invitation of one of the latter's correspondents in the Society – Whewell, for example.

Jacobi[14] was a well known and respected figure in Germany, a friend of Dirichlet, Bessel and his illustrious older contemporary Gauss. He received personal copies of the *Transactions* of the Cambridge Philosophical Society which, in 1839, recorded the receipt of the 'Kön. Ak. der Wissenschaft'. Jacobi was at Königsberg from

1826 to 1844. His work was known in Cambridge: Robert Murphy included him in acknowledgements in the preface to his book on electricity published in 1833.[15] This clear instance of a contact abroad may be the one case which gratified Green in indicating foreign recognition of his work, or it may be one of several which have not yet come to light. In either case it is a matter for regret that this recognition came so late, and that October 1839 saw the last of Green's publications, since he was not to live another two years.

The account of Green's life in Cambridge has focused attention on his later papers, but it should be recalled that what is now considered his greatest work, the Essay, was little known, either there or elsewhere. Only three instances of its existence in Cambridge are known. Caius College possesses a copy, presumably donated by Green. One copy was passed to the Cambridge Philosophical Society in 1832, and William Hopkins acquired three copies – a fact which will prove of some significance later in this narrative. One of these copies has survived. It is inscribed 'W Hopkins Esquire' in Green's handwriting, though without the usual addition of 'from the Author'.[16]

The Society's copy of the Essay also played its part in later events, since this was presumably the copy sent by Green in 1832 with his first paper on ellipsoids,[17] and read by Robert Murphy who was, it will be recalled, the rapporteur for the memoir. Murphy subsequently made a short but highly significant reference to it in a footnote to a paper on integrals, read by the Society in November 1833, and which later alerted William Thomson to the existence of the Essay. But if the Essay remained unknown during Green's lifetime, Green's papers did not. Jacobi provides a positive instance of a foreign contact on a personal basis. Nearer home, Matthew O'Brien, who was a Fellow at Caius with Green, referred in a paper to what Whittaker calls 'Green's great memoirs of 1837 and 1839, on the elastic solid theory', and adds 'he was obviously well acquainted with them'.[18]

Opinions vary as to when Green did the bulk of his mathematical work. H.G. Green maintains that his disappointing result in the Mathematical Tripos was the result of his continuing his own investigations while studying for his degree, and that it had also undermined his health, necessitating his departure from Cambridge.[19] Venn, who compiled the *Caius Biographical History*, states that his mathematical work 'was almost all done before graduation'.

Venn would have known Senior Fellows who had been in College in Green's time. The Reverend J.J. Smith, for example, had been Assistant Tutor when Green arrived in 1833. He would also have known Harvey Goodwin, who had been examined by Green. Venn was inclined to rely on hearsay in compiling his biographies,[20] but it is possible that his statement stems from discussions with those who had known Green.

It may be contended that Green had indeed done a considerable amount of work before he came to Cambridge. The importance of the Essay, in particular, might justify this contention. It developed ideas in electricity and magnetism which would later be taken up by Thomson and Stokes. It provided the term 'potential' and – most importantly – it produced Green's functions and Green's Theorem, powerful tools in the study of modern science.[21] On the other hand, the Cambridge memoirs have led to important advances in concepts applicable to modern developments in science and technology. It is still arguable that Green may have covered much of the ground before he came to Cambridge: the reading dates of the six papers – from 15 May 1837 to 20 May 1839 – might justify this view.

While Green was adding to his mathematical reputation through the publication of his memoirs in the Cambridge Society's *Transactions*, he was of course still living in Caius College, though except as a resident graduate he held no position or office. For this he had to wait for an election into a Fellowship. The likelihood of a Fellowship was a matter for long-term speculation by the ambitious undergraduate. This was advanced as a reason for Charles Babbage's migration in his first year, 1811, from Trinity,[22] one of the largest colleges, to St Peter's, only a 'Small College' – erroneously, as Babbage did not take the Senate House examination, and without Wrangler status no Fellowship would have been possible. Conversely, Thomson, in his first term at St Peter's in 1841, was already doubting his chances. A friend was thinking of moving to Queens' and he wondered whether, as the chances of a Fellowship at St Peter's were limited, he had better not also think of changing colleges.[23]

Each college had a separate foundation and evolved its own traditions and procedures, as indeed each had its own endowments and income. In 1592, a generation after John Caius had revised its statutes, the government of Caius College was vested in a Council comprising the Master and twelve Senior Fellows.[24] The 'Gesta' recorded their decisions and indicated their various responsibilities: care of buildings, management of rents, endowments and benefactions, as well as disciplinary and domestic concerns. The Senior Fellowships held by the Twelve were referred to in Green's time as 'of the Foundation' or 'upon the ancient foundation'. They carried a considerable stipend, since traditionally the Senior Fellows 'had divided the spoils between them for three hundred years'.[25] Later benefactions made possible 'Junior' Fellowships: the Frankland bequest provided six, the Perse five, Wortley two, Stokes (or Stokys) and Wendy one each. They carried smaller or negligible stipends. The Perse Fellowships, for historical reasons, had been much reduced, and by Green's time they were usually offered as an additional contribution to another appointment. It is impossible, from the College registers, to define the actual income received. The sums against Fellows' names vary considerably, since many would include fees for tuition, and fees from students could be substantial. In 1845 Thomson's father paid Hopkins £40 for vacation coaching and the final term's tuition.[26]

The Master and the twelve Senior Fellows of Caius College elected graduates to the Fellowships that fell vacant through departure, death or promotion. Since the stipends of the Junior Fellowships varied, it was possible to promote a man from one Fellowship to another that was better endowed. For example, Robert Murphy was promoted from a Frankland to the Stokes Fellowship, though he insisted that this should not prejudice his election to a Senior Fellowship. On one or two occasions during the year the 'Gesta' recorded a non-election, the Fellows presumably having failed to come to an agreement. Then, on 31 October 1839, George Green was elected to a Perse Fellowship worth £10 a year. The entry in the 'Gesta' for 31 October 1839 reads:

> In the Chapel: George Green B.A. after having subscribed in the Vice-Chancellor's book was sworn and admitted by the Master to a Perse Fellowship 'iis conditionibus quibus cautum est in annum probationis' in the presence of C. Eyres Registrary.

A few entries of elections contained this reservation of a year's probationary period.[27]

In Nottingham the news of Green's Fellowship may have been received with pleasure by those who were aware of what this promotion signified. It received a rather laconic mention in the *Nottingham Journal* a week later, on 8 November 1839:

> Cambridge – Mr. Green of this town has been elected Fellow of Gonville and Caius College Cambridge, of that University.[28]

For Jane Smith, the news must have aroused mixed feelings – pride in George Green's achievement, and satisfaction in his ambition of a Cambridge degree attained – but it also brought finality. Green had forsaken Nottingham for Cambridge, and his Fellowship precluded marriage. His will – and his daughter's epitaph many years later – would establish him as 'late of Snenton [*sic*] in the County of Nottingham and now of Caius College Cambridge, Fellow of such College'.

If the pleasures of his first year had been retained or were to be recaptured, Green now had more time for social contacts as well as for study, and some of these were provided by the College community. Caius, being a small college, would allow the normally reserved Green to make easier and closer contacts than would have been possible in the larger, Master-dominated establishment at Trinity, for example. True, he had not had till now access to the Fellows' Combination Room, but it may be assumed that as a middle-aged man Green had both the reputation and the means to entertain visitors and members of the College, and would on occasion have been a guest of the Fellows. So, in fact, it proved. A fortnight after Green had been elected a Fellow of the Cambridge Philosophical Society, the College Wager Book has an entry for 20 November: 'Whether Mr. Payne and Mr. Price are ejected from the Representation of the Borough of Cambridge'. Mr Smith gave one bottle, followed by – and each wagering a bottle – Mr Turnbull, Mr Stokes, Mr Pindar, Mr Irwin, and, inscribed in his characteristic hand, 'Geo. Green'. Another entry for 9 May 1838 suggests Green's presence again as guest of the Fellows.

Election to a Fellowship ensured a pleasurable existence and security of livelihood until a Fellow either took a church living or entered into matrimony.[29] The College Betting Book allows a glimpse into the social life of the Fellows' Combination Room. After dinner, when the port circulated, it seems that conversation ranged over politics and University affairs, and then, as the evening continued and the decanter passed from one to another, over a wider range of sometimes quite unlikely topics. To sustain the interest and perpetuate excitement, the long-established habit of the sporting Englishman – or, in the case of Robert Murphy, the sporting Irishman – asserted itself, and bets were laid. These were formally recorded in the College Wager Book. If one reads the entries from November 1839, the date of Green's election as Fellow, one finds that he waited until January before placing a bet – predictably, perhaps, on a mathematical topic:

> Mr. Tozer bets Mr. Stokes that the solution of Mr. Gaskin's problem about dice is not one eighth.
> lost by Mr. Stokes and posted.
> Mr. Green bets Mr. Stokes the same.
> lost by Mr. Stokes and posted.[30]

Significantly, perhaps, Green did not launch the bet, but comes in as a seconder. Later in the month he was one of six offering 'to give the room two bottles of wine, if the ministers resign during the present week or if the house express by a majority their want of confidence in them'. No fewer than twelve bottles of wine were lost to the Fellows on the outcome of this parliamentary crisis. This was, of course, a more serious bet than one as to whether a certain horse was the sire or grandsire of another, or how many eggs would remain unbroken when dropped to the ground, yet this last may not have been quite so frivolous as it seems. The laws of chance were a valid subject of mathematical study,[31] and Green's fellow participants in these bets were all mathematicians.[32] Apart from the two cautious bets already noted, Green's signature appears as witness on a couple of occasions, as for example when 'Mr. Paget bets Mr. O'Brien [an Irishman] that in Lewis's Topographical Dictionary of Ireland, no town is more than 50 miles from its true place' – lost by the rash Mr Paget (on what basis of proof, one wonders?). Green was again the witness when 'Mr. Tozer bets Mr. Stokes that there are not three counties whose longest dimension is longer than that of Kent.' This wager was also lost 'and posted', as Green records. The last entry was for 9 March 1840, when 'Mr. Green' appeared as one of four subscribers to a bet.

This seems to be the extent of Green's active participation in Common Room wagers during the six months or so he was a member. His modest contribution contrasts with that of Robert Murphy. During October and November 1832 Murphy laid ten bets, and during the following year he was involved as participant or witness in no fewer than twenty-one of the sixty bets recorded. In 1834 he placed twenty-three bets and in 1835 twenty-four, the last on 22 December. In each year the dates of the entries largely cover the College terms. The subjects of these and others in subsequent years range from political wagers, such as 'Mr. Stokes bets Mr. Murphy that there will be at least forty-five members returned from Parliament at the next election', to wilder speculations: 'Mr. Murphy bets Mr. Stokes [yet another Irishman] that Napoleon is said to have sacrificed the life of his wife rather than that of his son', or again, 'Mr. Paget bets Mr. Murphy that Abraham took to himself a wife after the death of Sarah'. Such bets, 'lost and posted' by one or the other, provided Murphy and his friends with a constant supply of bottles of wine. There are wagers involving six or twelve bottles of wine, but seven Fellows, including Murphy, gave six bottles each to celebrate the achievement of Mr Ellice of Caius as Senior Wrangler in 1833.

Robert Murphy enters the Green story at various points, so it might be appropriate to bring him a little more into focus. One of seven children of an Irish shoemaker, he showed an early precocity in mathematics. Local philanthropy enabled him to continue his education in his home area until the age of seventeen, and then to go to Cambridge. A sponsor sent some of his mathematical papers to Robert Woodhouse who, much impressed, offered, if Murphy arrived in Cambridge with £60–80, to see him through his course.[33] (Murphy's background of poverty and lack of family financial support should not be forgotten in the light of what follows.[34]) Murphy was Third Wrangler in 1829 and was immediately elected into a Perse Fellowship, at the age of twenty-three. He was also awarded the 'Schuldean Inkstand' as a mark of academic distinction.[35] In the course of the next nine years Murphy was promoted in 1834 to a Frankland Fellowship, worth £25 per half-year, and then in November 1838 to a Stokes Fellowship, worth £95 per half-year. These were Junior Fellowships. As the coveted Fellowships were those of the twelve Senior Fellows, this probably accounts for a meticulously worded resolution, presumably occasioned by an eloquent plea from the fiery Mr Murphy, which reads as follows:

> Apr 5 1839 'Resolved unanimously by the President and Fellows present that this election of Mr. Murphy into the Fellowship of Mr. Stokys's Foundation shall in no degree prejudice the question of his election at any future time into a Senior Fellowship.'

At the beginning of each academic year, in October, the College offices and duties were allocated: the Bursar, Registrar and Key Keeper, Steward and Key Keeper, Dean and Junior Dean (with the duty to 'conduct' – that is, to read the prayers in chapel), Catechist, Salarist, Hebrew Lecturer, Greek Lecturer, Praelector Rhetorician and Librarian, the last two appointed by the Master. The allocation of these was preceded by the reading of 'the Statutes of the Realm and college concerning elections'. Some of these offices carried a modest stipend. In 1835, for example, it was agreed to pay the Hebrew Lecturer – who, incidentally, was Robert Murphy – £10 a year. It was not an onerous appointment in view of the stipulation, announced by the Senior Fellows that the lecturer must give at least *six* lectures a year.

It seems that Murphy's impecunious background was not forgotten by the Fellows. In 1829, on his election to a Fellowship, he was also appointed Librarian, and in October 1831 he was made Junior Dean.[36] As Dean he was responsible for College discipline as well as services in the Chapel. In October 1832 he was again made Junior Dean, and also Greek Lecturer. The following December, however, the entry for the Fellows' Meeting records: 'That Mr. Thurtell be appointed Dean and conduct'. In his biography (1854) De Morgan wrote:

> He [Murphy] gradually fell into dissipated habits, and in December, 1832, left Cambridge, with his fellowship under sequestration for the benefit of his creditors. There is much excuse for a very young man, brought up in penury and pushed by the force of early talent into a situation in which ample command of money is accompanied by even more than proportionate exposure to temptation. His college admitted the excuse to the fullest extent: and though it could not tolerate the continued residence of an officer who had shown such an example, yet it was understood that his ultimate promotion to one of the more valuable fellowships would take place, on the amendment of his excesses.

The tragedy lies in the last phrase. The *Caius Biographical History* states uncompromisingly: 'A very brilliant mathematician but utterly wrecked by his dissipated habits'. The College Wager Book suggests evidence for these. Other entries in the 'Gesta' reveal Murphy's inability to handle money. Even as an undergraduate in December 1827, he was advanced 'thirty pounds out of the Perse and forty pounds out of the College Fund'. The following December the Senior Fellows 'agreed to advance Mr. Murphy a loan of fifty pounds'. At the same meeting he 'passed to' his BA degree. In December 1832 it was agreed 'that sixty pounds be allowed to Mr. Murphy for the present year'. It is at this same meeting, however, that we read of the resolution 'That Mr. Thurtell be appointed Dean and conduct'.

This is the event, so discreetly recorded, which is apparently the basis for De Morgan's statement that Murphy left Cambridge in December 1832. This was not so. The evidence from the College Wager Book shows Murphy in almost continual residence until late December 1835, and the 'Gesta' records his appointment as Hebrew Lecturer in 1833, 1834, and 1835. It has been assumed by mathematical historians, in the light of De Morgan's statement, that Green and Murphy never met, but this assumption does not accord with the facts. What is clear from the 'Gesta' is that at the Fellows' meeting in December 1832, Murphy, because of his debts and excesses, was relieved of his duties as Dean, and hence of all disciplinary responsibility for students. De Morgan is writing over twenty years after the incident and eleven years after Murphy's death. Whether the error came about through the passage of time, the exercise of discretion on the part of De Morgan, or Murphy's own presentation of the facts will never be known, but its correction may be significant for mathematical historians studying the work of Green and Murphy.

So Murphy's name – and none is more frequently found at this period in the 'Gesta' or the Wager Book – continues to appear until 1836. He ceased thereafter to be a resident Fellow, but paid visits from time to time – for example, to attend for his election to the Stokes Fellowship in 1838 and to establish his eligibility for a Senior Fellowship in 1839. There is, however, no further record of any election to a College office. But there is one final resolution recorded in the 'Gesta' on 22 January 1836: 'to advance £20.16.9 to Mr. Murphy as a loan for the payment of the Porter, Bedmaker, Laundress, Shoecleaner and Milkman'.

Murphy was a member of the Cambridge Philosophical Society; the Minutes record the reading of a number of his memoirs on integrals between 1830 and 1835.[37] In 1832 and again in 1835 he was elected a member of the Council. At some date after January 1836 Murphy moved to London, where Augustus De Morgan sponsored his writing a book, *A Treatise on the Theory of Algebraic Equations*, which was published in London in 1839 by the Society for the Diffusion of Useful Knowledge. Murphy also published two papers in the *Transactions* of the Royal Society, of which he was elected Fellow in 1834; Whewell and Peacock, amongst others, acted as sponsors. Finally, in 1838, he was appointed Examiner in Mathematics and Natural Philosophy to the University of London. Despite the earnest support of De Morgan, however, Murphy's weakness prevailed. Burdened as he was with continual debts, his health ruined by his excesses, he died in London in May 1843, at the age of thirty-seven. His portrait still hangs in the Fellows' Parlour, painted in 1828 by Dr John Woodhouse, Fellow of Caius and brother of Robert; it depicts a lean, dark young man with an arresting gaze.

Several tributes have been paid to Murphy. De Morgan, writing in 1854, held that 'He had a true genius for mathematical invention';[38] modern writers place him firmly in the company of Airy, Babbage, De Morgan, Green, Herschel and Whewell:[39]

> Murphy's papers on definite integrals are written at a very high level of mathematical sophistication. Further they reveal a high degree of planning and organisation, they are tightly written for the time, and they contain developments of substance far beyond those of his contemporaries.[40]

Mention has already been made of his university textbook on electricity, published in Cambridge in 1833, one of the very few works on this subject at that time. In the light of these achievements, Murphy may be considered the only mathematician of stature in Caius, and possibly in Cambridge, with whom Green could readily associate. If there is a matter of regret with regard to their association, it is that during the time of Murphy's residence in Cambridge, up to the beginning of 1836, Green was still involved with examination work. But on the later occasions when Murphy's presence is recorded in the 'Gesta', the College Wager Book or the Philosophical Society's meetings, it would seem unlikely that they did not at times discuss matters mathematical.

Beside the colourful and ultimately tragic figure of Robert Murphy, Green's other associates at Caius make a less striking impression as mathematicians. The most notable was Matthew O'Brien. Born in Ireland in 1814, he was Third Wrangler in 1838 and more than twenty years younger than Green (Murphy was fifteen years younger). O'Brien too died young, at forty-one, after ten years as Professor of Natural Philosophy and Astronomy at King's College London, and also as Lecturer at the Royal Military Academy at Woolwich. In 1842 he read the paper to the Cambridge Philosophical Society in which he made reference to Green's paper on wave motion; in the same year he published an *Elementary Treatise on Calculus*, followed two years later by a *Treatise on Plane Geometry*. The third Irishman, William Haughton Stokes, was elder brother to the great George Gabriel Stokes, friend and colleague of William Thomson and one of the outstanding figures of nineteenth century science. One person who became an established Cambridge figure was George Edward Paget. He perpetuated a Caius tradition by becoming a distinguished physician: he was Regius Professor of Physic for twenty years – from 1872 to 1892 – and he was knighted in 1885.

One possible rival to Green as a mathematician in Cambridge was the Second Wrangler to his Fourth, James Joseph Sylvester. He was one of the celebrated Johnians and unlikely to have been a close associate of Green's. In any case, he left Cambridge after taking the Mathematical Tripos in 1837, since as a Jew he was denied a degree on religious grounds. He, however, was blessed with a long life and thus, alone of Green's contemporaries under review, his mathematical promise came to full fruition. Born in 1814, he was at fifteen a student at University College London, under the Professor of Mathematics, Augustus De Morgan. De Morgan had been Fourth Wrangler in 1827, but he left Cambridge for London, since as a Unitarian he was debarred from a Fellowship. Sylvester entered St John's and, having taken the Tripos, returned to University College, now as a colleague of De Morgan, before taking the Chair of Mathematics at Woolwich. He was made a Fellow of the Royal Society in 1839, at the age of twenty-five. After an initial visit to the United States in 1836, he returned in 1876 and was instrumental in helping to set up the new Johns Hopkins University in Maryland. Finally, in 1883, all religious bans at Oxford and Cambridge having been lifted,[41] he returned to England to take up the Savilian Chair of Mathematics at Oxford.

The very real isolation of Green's intellectual life in Nottingham was self-evident – at least until he met Bromhead. It could be that he felt, in a sense, somewhat similarly isolated in his later years in Cambridge. His early vision of Cambridge life had been through Bromhead's eyes. Bromhead had been a young undergraduate, fortunate in meeting up with a small group of contemporaries of undoubted brilliance, with eager, enquiring minds and a common enthusiasm for French mathematics. A generation later, the situation for Green was quite different. He was an older man – older, indeed, than his nominal academic superiors, the Tutors and Fellows of the College. Standing 'head and shoulders above all his contemporaries in and outside the University',[42] he was perhaps almost as intellectually isolated as he had been in the 1820s, and apart possibly from Murphy, with no access to a mathematical intellect to match his own. Perhaps he found companionship in the company of Hopkins, but Whewell was now Professor of Moral Philosophy and was soon to take possession of the Master's Lodge at Trinity; Airy was Astronomer Royal at Greenwich, Peacock was Dean of Ely Cathedral; and Sylvester was well out of the picture.

The formal activities of Fellows varied, and some have already been mentioned, including holder of a College office, a lecturer or examiner or, if they were fortunate, tutor to private students. Of Green's responsibilities as Fellow there cannot be much to say, since the final tragedy of his life is that having at last obtained his Fellowship at the age of forty-six, he stayed at Caius for only two terms. Sometime in the spring of 1840, presumably at the end of the Michaelmas term, he went back to Nottingham, and did not return. In the words of William Tomlin:

> he returned indisposed after enjoying many years of excellent health to Sneinton, Alas! with the opinion he should never recover from his illness and which became verified in little more than a year's time by his decease on the 31st May 1841.[43]

When Green died, the cause of death was given as influenza. This does not accord with a long-term premonition of death, unless it fell as the *coup de grace* to end a life that was already undermined by some fatal condition. Green had spent a large part of his life as a miller. Working for long periods in the flour- and dust-laden atmosphere of the mill, he may in fact have contracted 'miller's disease', a complaint analogous to the silicosis suffered by miners breathing in coal dust, and a condition which declared itself in early middle age. It is quite probable therefore that Green fell prey to a pulmonary complaint, undetected by a doctor, who ascribed the cause to influenza.

There have, however, been suggestions from two sources that Green had succumbed to alcohol. Harvey Goodwin, who was Senior Wrangler in January 1840, the year of Green's departure from Cambridge, wrote in his entry in the *Dictionary of National Biography*: 'I was twice examined by Green. He set the problem paper in two of my college exams . . . he never assisted as far as I know in lectures. This might possibly be owing to his habits of life. His manner in the examination room was gentle and pleasant.' One commentator on this passage was uncompromising. In 1920, the President of St John's College referred to this 'allusion to Green's "habits" which I have always understood to mean drunkenness in plain English'.[44] Some fifty years after Harvey Goodwin's personal contact with Green, Felix Klein wrote in his *Entwicklung der Mathematik der Neunzehn Jahrhundert*: 'Unfortunately the circumstances which led to his talent being discovered and brought to light were not beneficial for his health; after becoming established in Cambridge, he succumbed to alcohol.'[45]

On one of his visits to the United States, Klein met James Joseph Sylvester at Johns Hopkins University. When Sylvester was due to leave his American post in 1883 to take up the Savilian Chair of Mathematics at Oxford, he suggested that Klein should apply to be appointed in his place.[46] If Klein was already interested in writing his account of nineteenth-century mathematics and mathematicians, he would have found Sylvester, with his long contact in Cambridge, London and Woolwich, a rich source of information. Sylvester had taken the Mathematical Tripos with Green in 1837, and he had worked closely in London with Augustus De Morgan, patron and biographer of Robert Murphy, a Caius contemporary of Green. The Caius College Wager Book reveals one aspect of the Fellows' social life: it provided a regular source of alcohol, and this finally proved fatal to Robert Murphy.

The quotations above prove nothing, but they do indicate the separate sources for the rumours concerning Green. These are strongly refuted by a number of people, who maintain that the attribution of alcoholism is slander, or an exaggeration, since it was an established nineteenth century custom of academics to quaff quantities of wine or beer during their lengthy discussions. The strongest refutation lies in the fact that Green published some of his most brilliant work during his last year in Cambridge.[47]

In the absence of a definitive verdict – which is now unlikely to be forthcoming – the question must remain open, but there are still one or two question marks. One is over the silence and apparent lack of knowledge on the part of Caius College after Green's death. Another is over the later part of Bromhead's letter of 1845. Having mentioned the letter of introduction addressed to 'some of the most distinguished characters of the University', he concludes abruptly:

> So much for my knowledge of poor Green, but I have written to a gentleman at Nottingham, who perhaps may supply further particulars and you will in the meantime
>
> believe me
> very truly yours
> E.F. Bromhead.[48]

One cannot escape the feeling that Bromhead is trying to distance himself from Green. He would have known of Green's achievements in Cambridge, the Tripos result, the eight papers in the *Transactions of the Cambridge Philosophical Society*, and the Fellowship. William Thomson's enquiry – to which Bromhead's letter was a reply – had concerned Green's pre-Cambridge life, and the background to the Essay. Thus Bromhead could have written of Green in more appreciative terms, lamenting an early death which cut short so much promise. Instead, we have the rather dismissive and patronizing 'So much for my knowledge of poor Green'. Thomson's letter had been forwarded to Bromhead by the Reverend J.J. Smith, of Caius College, and Smith was Assistant Tutor at the time of Green's admission in 1833. Had Green been so reticent during his time in Cambridge that after six years they had no more information than the bare, official facts? Or was there some need for discretion? Certainly the need for discretion was keenly felt in later years. In the 1920s Joseph Larmor was urging silence both on this point[49] and also on the question of Green's celibacy, following the recent death in Sneinton of Green's youngest daughter, Clara. In these depressing circumstances judgement must be withheld, though regret and speculation are allowable as to what George Green would have accomplished further, had he lived.

As it was, the ailing Green cannot have known anything of the extent of his achievement. Only forty-five years old, his mind fixed on death, and with no possible inkling of his posthumous reputation, he went home to the faithful and steadfast Jane Smith. Their seventh and last child, Clara, was born some weeks after his return, on 3 May 1840.

Chapter 9

George Green's Family

It was Jane Smith who reported George Green's death. This took place in the house she was occupying, at 3 Notintone Place, Sneinton, and she was present when he died on 31 May 1841.[1] Thus it seems fairly conclusive that Green, returning from Cambridge in failing health, chose to join Jane Smith at the last. This is the only recorded occasion of the two living under the same roof.

Notintone Place was a pleasant, tree-lined cul-de-sac, nearly opposite the parish church of St Stephen. It was demolished a few years ago to make way for the Salvation Army Community Centre. No. 12 and the adjacent houses on either side have been retained as a museum, since No. 12 is the birthplace of William Booth and faces the presumed site of No. 3. William Booth was born in 1829, though the family moved away from Nottingham when he was a child. It is some coincidence however that the birth of one and the death of the other of two of Nottingham's famous sons should have taken place in such close proximity. Entry to No. 12 gives us an idea of the house where Jane Smith and the children were living. The front door opens into the small front room. A door in line with the front door leads into a short passage, past the foot of a flight of steep stairs, to the kitchen at the back. On the two upper floors there are likewise a front and a back room. Thus the house, one room wide and two rooms deep, has six rooms in all, in which the mathematician, Jane Smith and seven children were living.

Three Nottingham papers recorded Green's death: 'At Sneinton on Monday evening the 31st ult. George Green, Esq., B.A., Fellow of Gonville and Caius College, Cambridge'. A few days later a slightly longer announcement appeared in the *Nottingham Review*, possibly written by the editor, Robert Goodacre, son of Green's old schoolmaster.

> In our obituary of last week, the death of Mr. Green, a mathematician, was announced; we believe he was the son of a miller, residing near to Nottingham, but having a taste for study, he applied his gifted mind to the science of mathematics, in which he made a rapid progress. In Sir Edward Ffrench Bromhead, Bart., he found a warm friend and to his influence owed much, while studying at Cambridge. Had his life been prolonged, he might have stood eminently high as a mathematician.[2]

This last sentence reveals in a few words the extent to which Green's reputation is entirely posthumous. Not only was Cambridge silent on the subject of his death, but in his own town and locality he was – as is evident – almost entirely unknown. He is not mentioned in Wylie's *Old and New Nottingham* (1853), which lists a considerable number of citizenry past and present. Nor is he mentioned initially in Mellors' *Old Nottingham Suburbs* (1914), though in the Addendum he states:[3] 'Principal Heaton calls my attention to a remarkable man, notice of whom ought to have appeared in the Sneinton paper',[3] and proceeds to give a belated and apologetic dozen lines. These seem to be the only local notice on George Green, until Dr Edith Mary Becket of University College Nottingham, the first to attempt an enquiry into the life of the mathematician, published her findings in the *Transactions of the Thoroton Society* in 1921.[4]

George Green was buried on 4 June 1841 in the same grave as his parents in St Stephen's churchyard. The inscription reads:

Also George Green, Esq., B.A.
Fellow of Caius College, Cambridge,
son of the above George and
Sarah Green who departed this
life on the 31st day of May 1841
Aged 47 years.

The name and titles of George Green were already known to the incumbent, the Reverend W.H. Wyatt, since in January 1840, perhaps to commemorate his new status as Fellow, Green had subscribed £10 to the newly rebuilt church at Sneinton,[5] where either he – or, more probably, his father – had been churchwarden. It was not normally the custom for women members of the family to be present at a funeral.[6] In all probability Jane Smith was not there, nor possibly was Ann Tomlin or her women relatives. More probable mourners were Green's cousin and brother-in-law William Tomlin, his maternal cousin Lewis Hartwell, and the Reverend Samuel Lund, friend and neighbour – both the latter being also executors of his will. Jane Smith's presence would probably have been an embarrassment; in any case she had her domestic responsibilities in Notintone Place, as there were now seven children. Jane, the eldest, was sixteen; Mary Ann thirteen; George eleven; John nine; Catherine seven; Elizabeth four; and Clara nearly thirteen months. Clara was born on 3 May 1840,[7] possibly just after her father returned from Cambridge in the spring of that year.

Green made his will that summer, in July, and added a codicil concerning the household goods the following January.[8] Significantly, he describes himself as 'late of Sneinton and now of Caius College, Cambridge, Fellow of such College'. To Jane Smith he left seven houses in Charles Street, inherited from his father, property in Wheatsheaf Yard and Warser Gate, and other real estate in Nottingham. On her death the property was to pass to his sister, Ann Tomlin. The house in Notintone Place is not mentioned, though in the codicil household goods, furniture, plates and linen are left to Jane Smith. The conclusion is that it was rented,[9] with Jane paying the rent from the income presumably supplied by Green for herself and her children, and later from that derived from the property he left her.

It appears unlikely that Green's prosperous relatives would have approved of his liaison with Jane Smith, and this could have been a factor in discouraging the ailing George from finally legitimizing his offspring. For example, his aunt, Mrs Ann Whittaker, *née* Butler, served a notice on 'Mrs. Jane Smith of Sneinton' in 1852 for an alleged breach of property rights, apparently based on a misunderstanding.[10] 'Mrs Jane Smith' – in the legal notice, at least – is not accorded the courtesy title of Mrs Jane Green by which she was generally known.

The question arises: Why did Green never marry Jane Smith, either earlier or even at the end, on his deathbed? Local tradition had it that George Green Senior was against the marriage. The Reverend F.C. Finch, vicar of St Alban's Sneinton, reported in a letter to Sir Joseph Larmor, written in 1919:[11]

> Today I have had a chat with an old Sneintonian who knows a great deal about the Green family. George Green, the great mathematician, was never married. The reason was that his father, the miller, did not approve of the woman who was the mother of George's seven children, declaring that if George married her he should never have a penny of his money.

This 'old Sneintonian' appears to have had a sister who was at school with Clara Green. This is a reference to events of more than eighty years earlier, but Green's situation – if it is true – was not unknown, given the parental authority and filial dependence and respect of the period.[12] An additional factor is that after his father's death came the possibility of Cambridge and ultimately a Fellowship, for which celibacy was required. And once he had the Fellowship George Green did not, apparently, wish to relinquish it – even on his deathbed. Such, perhaps, is the significance of his self-designation on his will, and the recording of his Fellowship in his epitaph.

Although he did not give his children his name and confer legitimacy on them, Green provided for them well, leaving them the Sneinton property acquired by his father. The other bequests were a £10 annuity to his uncle, Robert Green of Saxondale, whom he may have visited on his journeys to Thurlby; and £50 to each of his executors: Lewis Hartwell, Samuel Lund and George Eddowes, a solicitor. Green identified each of the seven children by name and date of baptism, though he never acknowledged their paternity. He seems to have had a regard for the basic patrimony, in view of the elaborate provision, made in the event of any one child's death, for the resulting share to be divided equally between the survivors. As all were minors at his death, arrangements were also made for trusteeship. The first to die was John, in 1852, at the age of twenty. He had been left the mill, so on his death each of his siblings owned one-sixth of it. This event must have heralded so many future complications that the first of several deeds was drawn up relieving the estate of the necessity for further division.[13]

The Sneinton property comprised the mill with the yard, outbuildings, workman's and foreman's cottages, the family house, and the surrounding area known as Green's Gardens. The different portions allotted to the children were meticulously delineated and mapped. The mill was left to John, the house to Clara. The rest was divided between the other five children. The bequests to John and Clara call for comment. The mill was not left to the older son, George, but to the younger son, John, then aged nine. This is strange at a time when the first born son normally took over the family business. Instead, George received nearly half of Green's Gardens, the remainder being divided between four of his sisters. A possible explanation – in view of the fact that George became a schoolmaster and subsequently took a mathematics degree at St John's College Cambridge – is that his father had recognized something of himself in his son, and had respected a nature more inclined to study than to the practical life of a miller. Certainly the mathematician sent his son to Goodacre's Academy,[14] by then sited on Standard Hill. The other rather strange bequest was of the family house to Clara, who was then only fourteen months old.

So when George Green died in 1841, he left substantial property, and for some years he had been accorded gentleman status. He also had wide family connections. The Greens and the Tomlins still farmed at Saxondale and Shelford, and the Butlers were firmly established as independent tradespeople in Nottingham.

Jane Butler, the mathematician's maternal aunt, had married Robert Hartwell, a pawnbroker in Long Row, on the opposite side of the Market Place to Angel Row, where stood Bromley House. They had two sons and six daughters. Their elder son, Lewis, later took over the shop and developed the clothing side of the business, though their younger son is listed in the town directory as 'Mr. Richard Hartwell, gentleman and proprietor'.[15]

William and Edmund Tomlin were also established as gentlemen. William became a particularly influential resident of the town, taking on duties such as a member of the Committee of the Vagrant Office of St Mary's Parish, or acting as municipal auditor.[16] George Green also had the assets to have established a similar standing in Nottingham. Unfortunately, instead of a prospering family like the Tomlins, in which Eliza Ann and Marion were known for their charity and good works, and their sons were recognized as gentlemen – William practised as a surgeon; Alfred exercised no profession, but lived as an eccentric and a man of leisure – George Green left a common-law wife and seven illegitimate children.

The respectable Tomlins, reflecting the moral outlook of their class and period, could only regard this orphaned family as at best unfortunate and at worst disgraceful. Furthermore, though their cousin and uncle had shown great promise as a mathematician and even achieved a Fellowship at Cambridge, his reputation in that direction was uncertain. William Tomlin may have been alerted to rising Cambridge interest on receiving the enquiry to which he responded in 1845, but thereafter he presumably heard no more. Nottingham was ignorant of the Cambridge publications – apart from one or two people such as Lewis Hartwell, to whom Green sent at least two papers, inscribed 'Lewis Hartwell Esquire from the Author'. But the Essay of 1828 was long forgotten; Green had resigned from the Bromley House Library eight years earlier; and his death at the age of forty-seven was unfortunate – because it had cut short his career, if for no other reason. Thus Green's legacy in Nottingham were Jane and the children; the Tomlins, and possibly others, preferred to ignore the consequences of an unfortunate liaison.

Jane Smith became known as Mrs Jane Green, and continued to reside at 3 Notintone Place for the next thirty years; at one time her neighbours included the Reverend Samuel May Lund. H.G. Green writes:

> The rents of the Nottingham property were sufficient to provide for the maintenance of the mathematician's widow in a quiet but sufficient style – indeed local report has it that to her death in 1877, she continued to dine in the evening, in itself a distinction in a provincial town of those days.[17]

After the deaths of Mary Ann and Catherine, she lived with Elizabeth and Clara, moving for the last year of her life to Belvoir Hill, where she died in 1877, aged seventy-five. Her grave, in which Catherine and her son George had already been buried, lies adjacent to the Green family grave: she was buried as Jane Green. The writer has heard at first hand of the strong local disapproval at the interment of Jane, Catherine and George in this spot – 'being what they were', yet another local report states with what regularity the Green family went to church. The impression left by Jane Green is not, however, of a poor downtrodden woman, the hapless victim of family snobbery and prejudice, but of one who held her head high, and lived her life in self-respect and natural dignity. The lives of her adult children would show that they had been bred in a household of industry and virtue. The story of these children will in turn reveal the extent to which their father's name and reputation came to be forgotten in the town of his birth.

All George Green's seven children survived infancy, which is rather surprising in view of the high infant mortality at that time, but possibly due to the fact that they were virtually brought up in the country: Sneinton was outside the town boundary until 1845. Three died in their twenties: John at twenty, and six months before inheriting the mill; then Mary Ann and Catherine. Catherine was buried in a new grave beside her father's, but the site of Mary Ann's grave is unknown. The mathematician's surviving son, George, never married. After attending Standard Hill Academy, he embarked on a mathematical career. In the estate deed of 1852[18] which sought to simplify the provisions of the mathematician's will after John's death, George is described as 'mathematical tutor of Netherleigh House, Chester' and also as 'a bachelor . . . and reciting that he had adopted the name of George Green and that the other six children of the said Jane Smith were also generally known by the surname of Green'. At twenty-six he went to St John's College Cambridge as a sizar. The President of St John's, Dr R.F. Scott, wrote of him later:

> Young George was actually recommended by Rev. W.H. Wyatt, M.A., of Pembroke College, who was Perpetual Curate and Vicar of Sneinton. Mr. Wyatt knew that Green had certain abilities and in spite of his age suggested he should take a degree at Cambridge and recommended him to the College. He then came but either his powers were over-estimated or his age was against his going through the drudgery of the Classics then required by all.

This was the Reverend W.H. Wyatt who had rebuilt St Stephen's, Sneinton, in 1839, to which project the mathematician had contributed his £10. He was therefore a near neighbour of Jane Green: Notintone Place was only a hundred yards away. One recalls the account of the Senate House examination as George Green had experienced it in 1837, and of the intensive coaching the St John's students underwent. Perhaps young George Green could not cope with the rigour of the examination, and may not have had the means to pay for the necessary coaching. In the final examination, he was Third Senior Optime. Scott goes on to say he was 'disappointed with his degree as he hoped to be a Wrangler'.[19]

Dr Scott was writing to Dr Edith Mary Becket in 1920, when she was investigating the anomaly of George Green's Fellowship at Caius, for which celibacy was required, and the existence of the family he had fathered, since no proof of marriage had been found. Dr Scott had written to a few of young George's contemporaries at St John's who were still living:

> Only a few contemporaries are now alive. They recollected the man, older than most, very reserved and not seeking companionship. . . . There was a belief that he had been a schoolmaster before coming up. Some resented the suggestion, now I believe settled by you, that he was illegitimate. For the elder Green on account of his mathematical prowess, was held in great respect.

Dr Scott continues, quoting from a letter from a surviving student:

> I remember George Green perfectly. He was by far the most impressive in his appearance of all the freshmen of our year. He had a lofty and broad forehead and a grave and dignified manner. He must have been about thirty years of age and had been a Schoolmaster before he came up.[20]

Green was actually admitted a week before his twenty-sixth birthday; he took his BA in 1859, when he was thirty. Bishop Harvey Goodwin remembered the father's manner as 'gentle and pleasant': now we hear that the son had 'a grave and dignified manner'.

Young George's reserve, his 'not seeking companionship', may have been natural to him, and again, indicative of a facet of his father's character. It may equally have been caused by the awareness of the circumstances of his birth and a reluctance to establish any close personal relationship which might have occasioned revelation of the truth. He was aware that this was known to the College authorities, since admission to Cambridge at that time required the production of a baptismal certificate, and Green's certificate recorded his mother as Jane Smith: his father's name was not entered. The inference was obvious. There was scant risk of association, however, unless he declared himself, since there were a number of George Greens in Cambridge over these years. He had entered St John's College on the sponsorship of the Reverend W.H. Wyatt, though a legitimate son of George Green would readily have been granted a place at Caius. But because he was the illegitimate son of the mathematician, young George Green could not claim kinship with his father, who by now had an established reputation in Cambridge, so he had to live at St John's incognito.

Not so his young cousin Alfred Tomlin. Alfred, the youngest of William Tomlin's four children, was admitted at nineteen as a pensioner at Caius College, where – according to Venn's *Biographical History* – he was recognized as 'Nephew of George Green the mathematician'. This statement is omitted in the larger *Alumni Cantabrigienses*, although the younger George Green is uncompromisingly entered as 'Illegitimate son of George Green'. George and Alfred's time in Cambridge overlapped by a year. George took up residence in 1855 at St John's at one end of Trinity Street; Alfred, seven years younger, in 1858 at Caius, at the other end. Alfred achieved Wrangler status, if a not very distinguished placing as Twenty-fifth, in 1861, and stayed on at Caius until 1864. The Register records: 'No profession. Resided for many years at Sneinton. Since at Prestwich, Lancs. (1898)'.

George Green, as Dr Scott relates, was disappointed that he achieved only a second-class degree – the more so in view of the fact that his father had been ranked Fourth Wrangler and even his dilettante cousin had emerged a Wrangler, though he was possibly helped by coaching paid for by his father. So George Green quietly left Cambridge, a disappointed thirty-year-old. A possible clue to his subsequent career may lie in an advertisement in the Cork newspaper in which the obituary of George Boole appeared. The *Cork Advertiser* of 2 February 1865 carried a notice by the Reverend John A. Wall, Principal of Portalington School, who named as his 'First Mathematical Assistant G. Green Esq., AB, Ex-scholar of St. John's College, and senior optime of the U. of Cambridge'. George Green is a common name, but the academic qualifications are correct. George may have known of Bromhead's patronage of Boole through his father and established a connection with him; this may have gained him an appointment in Cork.[21] The one sure event which brought George Green's life to a close, however, is found in the *Nottingham Journal* in March 1870:

> Suicide of a Gentleman. On Tuesday evening an inquest was held by Dr. Lankeshen the Middlesex Coroner at the Norfolk Arms Tavern, Half Moon Crescent, Barnsbury Road, relative to the death of George Green, aged thirty-nine, who on Saturday was found dead with a pistol wound in his head at 18 Barnsbury Road. The body was identified by Mr. Orlando Hallott, residing at Nottingham. He was a cousin of the deceased and knew him to be a tutor to various colleges. He had not seen him since last Michaelmas holidays. Deceased was unmarried and an M.A. of Cambridge.[22] He had not been in ill-health and was not in any pecuniary difficulties. Witness believed that his cousin had committed suicide. Further evidence showed that at a quarter to eleven on Saturday morning last deceased was found dead on the hearth rug by the fire, at 18 Barnsbury Road, the pistol being found in his right hand. After some remarks by the Coroner, the jury returned a verdict of 'Suicide while in an unsound state of mind'.

The *Nottingham Journal* was dated 17 March, and George Green was buried the next day in the grave where his sister Catherine lay, in St Stephen's churchyard. Suicide, until this century, was a crime, that of *felo de se*, and at this time, like illegitimacy, it was thought to bring disgrace upon the family. Suicides, moreover, were not always accorded burial in consecrated ground, and it perhaps says something for Mrs Jane Green's standing that her son was given a Christian burial by the Reverend Canon Hutton, vicar of St Stephen's. Jane was then in her late sixties, living with her daughters Elizabeth and Clara. To whom could she turn when the news arrived from London that her son had committed suicide? Who would go to London to perform the necessary identification of the deceased and make arrangements for the body to be brought to Nottingham? She could not ask her neighbours, the Tomlins, though the younger William Tomlin was a surgeon; nor, apparently, the Hartwells, who in any case were in their seventies. Instead she was able to call on her own family. Orlando Hallott was the son of Ann, Jane's elder sister, who had married George Hallott, son of Orlando Hallott, and given her son his grandfather's unusual name. In 1869 Orlando Hallott was a 'carver, gilder, picture framer & looking glass manufacturer' at 36 Warser Gate.

George Green Junior had made his will in 1863 in favour of his sisters Elizabeth and Clara. Elizabeth – known as Bessie in the family – had sold her portion of land left by her father in 1858, when she came of age, for the quite considerable sum of £1,500.[23] In this way, perhaps, she aimed to offset her illegitimate status by acquiring a handsome dowry, though she did not in fact marry until she was forty-two. Her husband, Amos Piggott, described as 'Gentleman' and living in Retford, handed over to Clara his wife's one-sixth share of the mill on her death in 1891.

Jane, the eldest child, married Frederick Thomas Moth of Retford and gave birth to a daughter, Mary Jane, in 1846, when she was twenty-two; and to a son, George Green Moth, three years later. Jane, who had helped her mother to bring up the six younger children and managed the household during her mother's confinements, must have been relieved to have founded her own small household, and passed on her responsibilities to her sister Mary Ann, who was three years younger. So in moving to Retford, she left the Sneinton area. Moreover, widowed at some date, she married Stephen Pickernell at the church of St Mary Magdalene, St Pancras, London, and after her mother's death she appears to have had no contact with her sisters in Nottingham.[24] Ironically therefore, the only child of the seven to take Green's descendants into the third generation disappeared from the scene. One contact only was made in 1910, through a legal transaction, when George Green Moth sold his mother's share of the mill, which he had inherited, to Clara for £35.[25]

There remained Clara, sole survivor and representative of the Green family in Nottingham, and she was in no circumstances to establish the reputation of a father whom she had never known. Her mother had died in 1877, and Bessie had left to marry two years later, so from 1879 until her own death forty years later, Clara Green had no family in the town and no settled home. The family house, by then known as Belvoir Mount, was hers, but she needed the rent as income; H.G. Green quotes local information that she was a paying guest of her tenant from about 1905 to 1916. She was also reputed to have been a governess in the family of the Reverend W.H. Wyatt, who had recommended her brother to St John's College. Her movements are shown in the various mortgage deeds by which she sought to raise capital on her dwindling estate. She was the sole inheritor of the mathematician's legacy of 1841, and thus the legal owner of the family house, the mill, and Green's

Gardens, less the portion Elizabeth had sold in 1858. By the turn of the century, however, the mill was surrounded not by summer houses and carefully tended plots but by scrub and run-down allotments. As for the mill itself, it had long been derelict.

When old George Green built his 'brick wind cornmill', he was not to know that wind-powered mills would be worked by only one more generation of millers. James Watt's steam engine was already in use by 1818 for powering the previously wind-driven watermills, long used in the Fens for draining the land. It was only a matter of time before cornmills, entirely dependent on the vagaries of wind and weather and necessitating uncertain hours of labour, were also powered by steam. A number of mills had engines installed to provide either full or auxiliary power, before being abandoned in their turn for the later steam-powered roller mills. Green's Mill was not so converted. By mid century, few mills in the area were in fact working. In 1858 the Nottingham Date Book records the burning down of the last of the thirteen mills once turning on the Forest. The 1861 census gives William Oakland as tenant of Green's Mill and resident there,[26] possibly in the foreman's cottage. Ten years later he was working a post mill in Windmill Lane, presumably the one entitled 'Oakland's Mill' in the drawing by T.W. Hammond. The cause of his departure was a dispute about responsibility for repairs between tenant and owners. As this was not resolved, Oakland must have vacated Green's Mill some time during these ten years. It was never worked again, and gradually fell into disrepair. One practical reason for the decay of the mill, in addition to the economic decline of windmills, was the fact that the rent from her property constituted Clara's sole means of livelihood. This would be a strong reason for her not to live in the house itself, but an even stronger one for the gradual deterioration of the mill, since there was no accumulated capital to spend on major repairs.

In about 1916, Clara Green appears to have gone to live in a summer house on her land near the present Finsbury Avenue in Sneinton. Here she led the life of an eccentric and a recluse, lonely, proud and impecunious. Elderly Sneintonians have recounted to the writer how as small boys they had taunted her and thrown stones, and how she had chased them off with a broomstick. 'We'd no idea who she was', they said. 'Of course we feel sorry now for what we did, but you know what boys are!' An earlier account of the elderly but still unsociable Clara was given by Professor Granger of the University College to Sir Joseph Larmor in 1907, in which he described his attempt to interview Clara on the subject of her father. She fiercely repudiated any enquiries, demanding 'if information was wanted, why it was not sought before'. 'But Miss Green,' he went on to say, 'though eccentric, is a woman of breeding.' She also:

> especially mentioned Sir Edward Ffrench Bromhead of Thurlby as having helped her father. . . . Nottingham has a touch of the rebel and the individualist and Green's career seems less surprising to those who know his birthplace. . . . Miss Green perhaps inherits from her father her exaggerated independence of mind.[27]

Finally, at the age of seventy-eight, Clara Green fell ill and was taken to the Workhouse Infirmary at Bagthorpe, where she died some two weeks later. The Reverend F.C. Finch, vicar of St Alban's, Sneinton, wrote to Sir Joseph Larmor:

> Miss Clara Green, the youngest daughter of George died a few days ago: age 84 [*sic*]. There were four other daughters and two sons, all now dead. She was the owner of the Mill (long disused) and the adjoining allotment gardens: very eccentric. She lived not in the Mill House but in a summer house in one of the gardens. She was removed to the Infirmary about 10 days before her death.[28]

It is Finch who gives us the information that George never married Jane Smith because of his father's opposition and it is Finch, writing at the same time to Granger,[29] who maybe puts his finger on Clara's problem: 'This [the fact that her parents never married] partly accounts for Clara's oddness: she keenly felt the stigma of her illegitimacy, and had an idea that people looked down upon her'.

The stigma of illegitimacy was strongly felt in the nineteenth century and into the twentieth. Where adultery was on occasion overlooked, any resulting pregnancy was not, and the shame attaching to the parents' conduct was visited in no uncertain measure on the innocent but unfortunate offspring. So it is possible that the Green family were the victims of changing social opinion, as standards of sexual morality changed in the nineteenth century.[30] As the Victorian age progressed, the royal example of Victoria and Albert became increasingly adopted in the national mind as the pattern of family life. This, with the coming of Nonconformism and the upsurge of public and private morality, led to a keener censure of what came to be regarded as 'immoral' living. Thus Green may not have erred unduly in contemporary eyes in fathering his family, particularly in view of what was postulated earlier regarding his filial obedience or his Cambridge aspirations. It was more probable that his children, as adults in the second half of the century, acquired a stigma as social mores changed. So the fortunes of Jane Smith and her children must be reviewed in the light of changing social opinion. One must admire the dignity of Mrs Jane Green, living out her life in a small community, with or without contact with the family of the father of her children, ponder the reasons for her son George's suicide – had he wished to marry, for example, but had no name to offer a 'respectable' woman? – and regret the misanthropy shown by Clara in her old age, if it is attributable, as stated by Finch, to the circumstances of her birth.

The correspondence of Professor Granger and the Reverend J.C. Finch provides us with further information about Clara. Dr Granger says: '. . . talking of the Mill House, she said that as a child she would often stand at the gate and look up at the stars, watching the Great Bear above the mills sails'.[31] Another testimony of Clara's which has been much quoted is reported by Granger: that her father used the top floor of the mill 'immediately below the great wheel' as a study.[32] This story provides one romantic link between George Green the miller and George Green the mathematician, and it is an interesting story to investigate. George was about twenty-four when his father built the family house, and he and his parents went to live there shortly after his sister Ann's marriage in 1816. It is a sizeable house for a family of three with possibly a servant living in: two large front rooms, a third which might have been a living room or an office, and five bedrooms. It is reasonable to suppose that George would not only have had his own bedroom but possibly, had he wanted it, a second room as a study. George, as stated in Tomlin's letter, found his miller's duties 'irksome'. One wonders why, then, he should trek across the mill-yard, climb four flights of stairs to the top floor 'under the great wheel', where there is no light, and lay out books and papers in a cramped space with no writing area in a dust-laden atmosphere; and then, collecting all his books and papers together, trek all the way back again at a late hour, in the dark, to his house.

Yet there are two possible justifications for this reminiscence of Clara's. Old George Green's will had made reference to 'the new erected house and gardens now in the occupation of myself and William Smith (my miller)'.[33] This suggests that Smith, now possibly a widower, was a member of the Green household, taking greater responsibility for the mill in old Green's failing years. (In parenthesis, one might deduce here a further reason for the old man's obduracy about his son's marriage to Jane. The established relationship of boss and employee would have been upset, with old George and William, as fathers in-law, placed on an equal footing – a prospect that old Green might have found distasteful.) After Green Senior's death, it may have seemed natural for Jane and the three children then born to move in with her father and theirs. In these circumstances Green would possibly have preferred the cramped – but warm and quiet – conditions of the mill. Yet when he writes to Bromhead in May 1833 'I have let the house I now occupy',[34] presumably to capitalize his assets before going to Cambridge, this would have necessitated finding other accommodation for the Smith family.

An alternative explanation for what, on the surface, appears a rather unlikely story is that it did not refer – as has been assumed – to the years of Green's maturity, the famous Essay written in the shadow of the great brake wheel, but possibly to a different period of his life. One must remember that Clara was only a year old when her father died, so the story must have come from her mother. Thus a more likely time when this could have happened is in George's younger years. The mill was constructed in about 1807, when he was fourteen, but the house was not built until some ten years later. Until then the adolescent George, the only son helping his father, may have walked from Goose Gate in Nottingham over the fields to Sneinton to spend days at the mill. He may have stayed over some nights, maybe to avoid riots and disturbances in Nottingham – for example, the Luddite Riots of 1811 and 1812 – or possibly to save time and energy. In Sneinton he may have spent the night in the mill manager's cottage, if he did not doss down in the mill itself, and perhaps William Smith was already the manager at this time. In such circumstances the young George might have kept his books and papers at the mill and worked in a quiet corner on the top floor, and Jane, the miller's daughter, would have known of this. It could be from this period that their relationship developed, though their first child was not born till 1824, and these early romantic memories were passed on by Jane in later years to her youngest daughter.

A further piece of information about Clara Green dates from 1937, when the then vicar of St Stephen's reported to an enquirer that 'two neighbours burnt all the rubbish in the house after her death' and he had no doubt that the records of the father went up in the fire. As late as 1972 an elderly Sneintonian recalled clearing out the summer house and burning a lot of papers, including, apparently, those of the mathematician.[35] If Clara did have some of her father's mathematical papers in her possession at her death, she must have guarded them with great filial devotion. They would apparently have been left by Green in Notintone Place and kept by Jane until Clara took them over on her mother's death in 1877. Thereafter she would have taken them with her to all of the five or so different addresses mentioned in the mortgage deeds, until they ended up in the summer house in the garden, before finally being consigned to the flames. If this is indeed what happened to Green's papers, it accounts for the fact that there is no trace of his sources or methods of working, other than what is evident in his published work.

Clara's will was made out in favour of her sister Elizabeth, who had died twenty-eight years earlier, and it had never been altered. It was assumed that she was without kin, since George Green Moth and his sister, Mary Jane, had vanished from the family scene, though the former had signed the mortgage deed only nine years previously. So the Crown took over the sale of the mill, the family house and the surrounding land, all the estate of 'Clara Green otherwise Clara Green Smith . . . a spinster and a bastard',[36] but the proceeds of the sale were insufficient to pay her debts.[37] Until 1989, it was not known where Clara Green was buried; it was presumed that she was interred in a pauper's grave. By chance her grave was located in the Rock Cemetery on Mansfield Road, Nottingham, and even then it took some finding, hidden away in a corner against the street railings behind the Entrance Lodge. Seventy years after Clara's burial, it looked as forlorn and as uncared for as doubtless its occupant once did, the headstone askew and half buried in the ground. The epitaph reads:

In Memory of Clara Green
Daughter of the late George Green M.A.
of Caius College, Cambridge
and formerly of Sneinton
Died 6th March 1919 aged 78 years.

In the light of the complete neglect in Nottingham of the mathematician and his family since his death nearly eighty years earlier, there is a certain poignancy in reading this proud claim of kinship by a daughter to an unknown father and a forgotten genius – the more so since another eighty years had passed before this inscription was discovered and read with any awareness of its implications. The question arises: Who had it inscribed on Clara's headstone? There are no directions to do so in her will. There are echoes of her father's epitaph, which point to a Sneinton connection. It might possibly have been the Reverend F.C. Finch of St Alban's Church, which is barely a quarter of a mile away from St Stephen's. On the other hand, the inscription may have been suggested by Professor Granger, of the University College.

Clara had outlived all her Nottingham family by many years, and the name of Green disappeared with her. The Tomlins, too, were now dead, and their name also had largely disappeared from Sneinton. Their fortunes, however, had taken a spectacular turn, and are worth a brief look before the family fade from this story. William Tomlin and his wife Ann, the mathematician's sister, died within a month of each other in 1880 after sixty-four years of marriage, an event commemorated by the gift to St Stephen's Church of a handsome brass lectern, donated by their daughters. They are buried in an imposing vault beneath the east wall of St Stephen's Church. With their names is also inscribed that of their elder daughter, Eliza Ann, 'of Langley House, Prestwich, Lancs.', who died fifteen years later.

One might not have paid too much attention to this phrase if one had not wondered where Miss Marion was buried, and then recalled the entry in the *Caius Biographical History* which states that in 1898 Alfred Tomlin was residing in Prestwich, Lancashire. Prestwich is now part of Greater Manchester, and the site of Langley House is engulfed in the Langley Housing Estate. Langley House was built in 1860 at a cost of £10,000, and stood in some ten acres of ground. It merited a place in the *History and Traditions of Prestwich* (1905), when it was noted that 'The present tenant is Miss Marion Tomlin'.[38] The Misses Tomlin first subscribed to St Margaret's Church in Prestwich in 1891, and continued to do so thereafter. In 1895, when Eliza Ann died, her sister had two 'opus sectile'[39] panels placed either side of the east window in her memory and, five years later, a three-foot oak Poor Box and various donations in memory of her brother Alfred. In 1901, to celebrate the church's jubilee, she had spent £75 on a brass lectern of singular magnificence. On her death in December 1910, at the age of eighty, St Margaret's Parish Magazine devoted a column to her eulogies:

> Miss Tomlin endeared herself to all her neighbours by her gentle disposition and large-hearted benevolence; there was never any occasion to appeal to her to help in any good or charitable cause; she was always seeking opportunities, as God's steward, for doing good . . . her great delight was to be kind and helpful to others, especially to the poor. . . . She was a true Christian both in faith and works, ever animated by the desire of following as closely as she could in the footsteps of her Saviour.[40]

William Tomlin had been known as a man of property in Nottingham, and to have been involved in various financial transactions. His will reveals their extent. In addition to four houses in Belvoir Terrace and two in High Pavement in Nottingham, he had purchased land in the Meadows area, between the southern boundary of the town and the River Trent, before its development for housing. He also owned land in Lowdham, Caythorpe and Gunthorpe, along the Trent. Furthermore, he had invested widely in railway stock, leaving to his son William, who had in fact predeceased him by a year, preference shares and debentures in the Dublin and Belfast Junction Railway; to Eliza Ann, shares in the Midland Railway; and to Marion, shares in the Great Western, the Manchester, Sheffield and Lincolnshire Railway, and the Great Eastern Railway.

When Eliza Ann died in 1895, she left £62,000; fourteen years later her brother left some £26,000; and in 1910 her sister Marion left nearly £30,000; they had all lived on their income since their parents' death in 1880. Alfred, eccentric and effete, characteristically did not make a will. Eliza Ann's and

Marion's wills are impressive in their extent and degree of detail. There are bequests to numerous Tomlin cousins. Those who farmed in Shelford can be identified; the many others mentioned have proved elusive. There were also legacies to friends – including, surprisingly, one of £100 to 'Rev. Thomas Lund, chaplain to the Blind Asylum Liverpool, (or to son)'. Thomas William May Lund was the son of the Reverend Samuel May Lund, friend and executor of George Green. Thomas Lund was sixty-seven when Marion Tomlin died, and had been rector of St John the Evangelist in Cheetham, Manchester, from 1871 to 1884.

One wonders whether Thomas Lund provides the key to the mystery as to why three elderly siblings should leave their native town, where they had spent their lives in social prominence, and migrate to an area north of Manchester, where they apparently had no connections. Perhaps a visit to the rector in Cheetham in the 1880s persuaded the Misses Tomlin some years later to establish themselves as the chatelaines of Langley House. There they presided over a household of at least eight domestic staff.[41] There were legacies to these and to friends in Sneinton, and charitable gifts to the poor, including '£5 each to each of persons in Sneinton receiving 2/6 per month',[42] but none of Marion's £30,000 went to her maternal cousin. At this time Clara Green was a tenant in Belvoir Mount, as the family house was now known, bereft of family and saddled with debts, a resident of Sneinton who was not among the beneficiaries of her wealthy cousin. Her lonely death in Sneinton in 1919 appeared to justify the concluding paragraph of Dr Becket's paper on George Green:

> The extinction of the family of the great mathematician and the disappearence of his books and papers have finally destroyed the hope of discovering either the sources from which Green drew his early inspiration or the means by which his mathematical genius was brought to light.

But George Green's family was not extinct.[43] His eldest daughter, Jane, in her second marriage, had gone to London and died in 1900 as Mrs Stephen Pickernell, and was buried in Richmond-on-Thames parish churchyard. Her daughter Mary

Jane died the same year, but her son George Green Moth was still alive,[44] and it is his descendants who will feature in a later event and will carry the mathematician's line, if not his name, into the next century. The merit for tracing the long lost Jane Smith, Green's eldest child, goes to Dr J.M. Rollett[45] who, over a period of years, gradually unravelled a complicated story.

The nub of the complication was that George Green Moth married twice, and at an interval of nearly forty years. As it happened, Dr Rollett traced the descendants of the second marriage first, and was able to meet Mrs Eva Saunders, the eldest of George Green Moth's three daughters, in 1974. To his great disappointment, Mrs Saunders and her sister Mrs Carey (the youngest daughter, Mrs Voss, was in Canada) knew nothing of their connection with the mathematician. George Green Moth, a sixty-year-old widower, had married their mother and had never, apparently, spoken of his previous family. There was thus no possibility of collecting any family souvenirs or reminiscences. There was only a snapshot of George Green Moth and his family taken in about 1916. His two eldest daughters, however, expressed strong interest in their new-found forebear, and both came to Nottingham some years later to visit the mill. Mrs Saunders and her family were able to be present in 1985 when the mill was officially reopened. But Dr Rollett still persevered, and the week before the opening ceremony he announced that he had finally traced the earlier family.

George Green Moth had married his first wife at the age of twenty-two. They had a daughter, Clara Jane, and two sons. Their mother, Anne Coombes, died young, and the three children were brought up by their aunt, Mary Jane, the mathematician's granddaughter. Clara Jane Moth married Thomas Higgens and had twins, followed by two daughters, Helen and Margaret. Helen Mary married Herbert Hall and farmed in Norfolk. Her younger sister Margaret Clara became an Anglican nun. It was Mrs Mary Hall whom Dr Rollett contacted in 1985 – a lady already in her seventies who, through these family circumstances, had a number of stories to tell. She also had a photograph of her great-grandmother, the mathematician's eldest daughter. This photograph of Mrs Jane Moth-Pickernell shows her composed and dignified, clad in a handsome dark dress, with a cameo brooch at the neck, which was still in Mrs Hall's possession. Jane's son, George Green Moth, appeared to have inherited her long nose and high forehead, as did his daughter, Mrs Eva Saunders. One wonders whether Jane inherited these characteristics from her father, and whether there is an indication here of how George Green himself looked.

Mrs Hall's anecdotes, passed down through three generations by the women of the family, all concern this Jane. They bring George Green almost within breathing distance, as well as the hitherto elusive Jane Smith. It is evident that Jane, lace-dresser, common-law wife, and mother of seven children, was a woman of some character and spirit. She travelled from Nottingham to stay with her eldest daughter in Richmond and was remembered by her great-granddaughter, Clara Jane, Mrs Hall's mother, as a 'brisk and lively old lady'. Clara Jane was five when Jane Green died in 1877 at seventy-five, so the impression must have been a vivid one. As for the Jane of the photograph, she appears also to have been a girl of spirit.

In 1831, the year of the Reform Riots, she was a child of seven. Angered by the defeat of the Reform Bill,[46] a mob of angry townspeople surged to Colwick and attacked the Hall, home of the landowner, Mr George Musters, whose wife, Mary, had been a friend of Lord Byron.[47] On their way back through Sneinton, however, the rioters saw the mill as an immediate target. There was also the fact that George Green was a local landlord and an overseer of the poor of Sneinton. Overseers were often unpopular figures, since the position was not always honoured either by the holder of the office or by the recipients of his administrations. The mathematician vigorously defended his property by firing on the crowd, and his young daughter Jane helped him to reload his musket. The mob then tore up the railings in Notintone Place and, armed with these, continued on their way back to Nottingham. There they proceeded to set fire to the Castle,[48] the property of the hated Duke of Newcastle, their MP, who had voted against the Bill.

At the age of ten, Jane worked a sampler – in silks, not wool – which was also in Mrs Hall's possession, an indication of an ordered and disciplined childhood and its social standing, or pretensions. As we noted above, there is proof here that the mathematician's family had already taken his name:

Jesus permit thy holy name to stand
As the first effort of an infant's hand,
And as her fingers on this sampler move
Engage her tender heart to seek thy love.
With thy dear children lend her a part
And write thy name thyself upon her heart.
Jane Green marked this sampler aged ten years 1834.

Jane continued to show a vigorous and independent turn of mind. At one time she was engaged to a well-to-do young man 'who kept his own carriage'. When he arrived one day with a small gift for one of her younger sisters, Jane flew into a rage and returned his ring. The most intriguing story of all, however, is how, on the night of George Green's death, Jane the mother and Jane the daughter, now aged sixteen, were sleeping together; the mathematician's body was laid out in another room. But there, in his chair by the fireside, they saw George Green sitting, so clearly that they could see his fingernails. And there he stayed for a while. Terrified, the two women cowered beneath the bedclothes, the older Jane exhorting her daughter: 'Look again, Jennie! Look again, Jennie!' Mrs Hall stressed the quiet conviction with which this story was always recounted by both mother and daughter. A final, rather bizarre note in this recital of the Green family fortunes is that neither of George Green Moth's families knew of each other's existence. Apparently Moth, who had never referred to his grandfather in Nottingham, had never referred either to the existence of his first family. The result was that neither knew of the existence of the other until Dr Rollett made his enquiries.

Jane's story, fortunately, is a happier one than that of either her younger brother or her sister, and it is a relief, after George's suicide and Clara's wretched death, to find that Jane enjoyed – at least the second time round – a happy marriage. Mrs Hall reported that Stephen Pickernell was so admired and beloved by his step-granddaughter that she named her first son Stephen after him. Jane's story allows us to conclude the Green family history on an optimistic note. Far from being extinct, the mathematician's family has descendants in both Canada and Australia and, nearer home, two branches who each feel a strong sense of pride in the achievements of George Green and in his restored mill. But this is to anticipate. It will be necessary to return to the early mid-nineteenth century and match this story of the final reappearance of George Green's family with the establishment of his scientific work, before finally describing the restoration of his mill.

Chapter 10

William Thomson and the Rediscovery of the Essay of 1828

One may only conjecture how Green's reputation would stand today, had his Essay on electricity and magnetism not been rediscovered by William Thomson.[1] It would have rested on the papers published in the *Transactions of the Cambridge Philosophical Society*, and thus chiefly on Green's contribution to studies in hydrodynamics, sound and light. The circumstances of the private publication of the Essay in Nottingham – and particularly the fact that only three recipients are recorded in Cambridge – suggest that but for two fortuitous incidents, the Essay would probably have remained forgotten. The main theories contained in it were reinvented later by others, but Green would have lost his priority, his reputation would have been the less, and the scientific world might not have had the use of the eminently valuable Green's functions. Undoubtedly, too, Green's reputation would not have stood so high if the young William Thomson had not been so enthusiastic about his discovery of the Essay, and the older Lord Kelvin had not publicly revered him throughout a long life as a distinguished man of science.

William Thomson's diaries and letters are useful in giving a picture of Cambridge undergraduate life in the 1840s. Those same sources[2] are also valuable in relating the circumstances of the rediscovery of the Essay, nearly twenty years after its publication and four years after Green's death. Since early adolescence Thomson had kept a mathematical diary in which he recorded the progress of his studies and his ideas for future research. On his return to Scotland in the summer of 1844, during which he prepared for the Senate House examination the following January, he wrote:[3]

> Gourock, Sept 10, 1844 – I have long been entertaining a project of writing a series of essays on the mathl. theory of principles, and giving all the applications of the general theorems relative to attraction, which are of use in giving a comprehensive view of the subject. Though my plan is not quite settled yet, and I shall have no time to think upon it until after I have taken my degree, I commence writing down some of my ideas in a disjointed manner.

Five pages of a draft outline follow. Then he added on 27 January 1845, less than three weeks after taking the Tripos: 'I have just met with Green's memoir [i.e. the Essay of 1828] which I first saw referred to in Murphy's first memoir, on Definite Integrals with physical applications, which renders a separate thesis on electricity less necessary.'[4] Two days earlier, on 25 January,[5] the day before his departure for Paris with his friend and fellow student Hugh Blackburn,[6] Thomson wrote to his father: 'Yesterday I got some separated copies of various memoirs from Hopkins and among them a most valuable one by Green, with wh. I am greatly delighted'.[7]

A detailed account of how Thomson acquired this valuable memoir by Green is found in a letter written by the aged Kelvin in his eighties, a month before he died, in response to an enquiry from Sir Joseph Larmor, who succeeded George Gabriel Stokes as Lucasian Professor of Mathematics in 1903, and was interested in finding out what he could about Green's life:[8]

> Netherhall
> Largs
> Ayrshire
> Nov. 19 1907
>
> Dear Larmor
>
> Yours of yesterday received. Green was never a pupil of Hopkins.
>
> When I went up to Cambridge as a freshman,[9] I asked at all the book shops in Cambridge for Green's Essay on Electricity and Magnetism, and could hear nothing of it.
>
> The day before I left Cambridge for Paris after taking my

> degree, in Jan 1845, I met Hopkins on what I believe was then called the Senior Wrangler's walk and I told him I had enquired in vain for Green's Essay and had never been able to learn anything about it all the time I was an undergraduate. He said 'I have some copies of it.' He turned with me and took me to his house, and there, in his chief coaching room in which I had been day after day for two years, he found three copies of Green's Essay in his bookcase and gave them to me.
>
> I had only time that evening to look at some pages of it, which astonished me. Next day, if I remember right, on the top of a diligence on my way to Paris, I managed to read some more of it.

Here, then, are the two incidents which took place in January 1845 and, together, led to the rediscovery of Green's Essay. The first relates to Murphy's reference to the Essay in a footnote to his memoir on the inverse method of definite integrals, which was published in 1833 in the *Transactions of the Cambridge Philosophical Society*.[10] The reason for Murphy's reference to the Essay in a footnote is as follows. Murphy's paper was read in March 1832. Green's paper on ellipsoids, for which Murphy was rapporteur, was sent with a copy of the Essay in May, but Murphy did not see it until after the long vacation, as Whewell explained in his letter to Bromhead of 14 November 1832. The comment on the paper was: 'there is part of the work which . . . coincides in some measure with what Poisson has done and with what Mr. Green himself has done'.[11] This last phrase undoubtedly refers to the Essay. So Murphy, having read the Essay only in November, could not amend his own paper on integrals, since it had already been read, so all he could do was add the following footnote:

> The electrical action in the third section, is measured by the tension of the fluid which *would* be produced by an infinitely thin rod, communicating with the electrical body, by the attraction or repulsion of the matter; it is what Mr. Green, of Nottingham, in his ingenious Essay on this subject, has denominated the Potential Function.[12]

The second incident was Thomson's chance meeting with Hopkins on the eve of his departure for Paris. If further proof were needed that electricity was not part of the Tripos syllabus, it is here.[13] The expensive tuition dispensed by Hopkins, resulting in his record number of Senior Wranglers, presumably concentrated on essential matters – namely, the subjects of the Senate House examination. In two years of tuition the topic of electricity had not, presumably, been discussed, otherwise it seems incredible that Hopkins would not have shown one of his most brilliant students Green's Essay.

An immediate consequence of Thomson's discovery of the Essay was its introduction to continental mathematicians, since his imminent departure for Paris was yet another link in this chain of coincidences. Thomson took the Senate House examination in January 1845 and spent the rest of the academic year in Paris, pending a hoped-for and confidently expected election into a Fellowship at St Peter's. He was at home in both France and Germany, and read French and German, since Professor James Thomson had taken his six children to both countries in 1839 and 1840.

Thomson and his father had a definite purpose in mind in William's visit to Paris. It was James Thomson's hope that his son would be appointed to the Chair of Natural Philosophy in Glasgow on the probable decease of the then incumbent, whose state of health had called for frequent comment in the correspondence between father and son. The holder of a Scottish Chair of Natural Philosophy was expected not only to give lectures but also, in some courses, to perform experiments.[14] Professor Thomson, in advocating his son's candidature for the Glasgow Chair, was anxious for him not to appear 'merely an x-plus-y man' – which in effect was what William basically was, since Cambridge study of mathematics was purely theoretical. So William was to spend four months in France to gain experience in performing scientific experiments, as well as making acquaintance with some of the foremost mathematicians then congregated in Paris. He took with him numerous letters of introduction. One was for Victor Regnault, Professor at the Collège de France. Thomson helped him with experiments in his *cabinet de physique*, gaining both experience and a testimonial.

So on his way to Paris, the twenty-one-year-old Thomson started his avid perusal of Green's Essay. He continued his account of events in his letter to Larmor of 1907[15]:

> Two days after that [i.e. his arrival in Paris], I called on several members of the Academy to whom I had introduction, among them Liouville and Sturm[16] bringing Green's Essay with me. I found Liouville at home and showed him Green's Essay, to which he gave great attention. I did not find Sturm at home but I left a card. Late in the evening, when I was sitting with my Cambridge comrade Blackburn, at our wood fire in 31, Rue Monsieur le Prince, we heard a knock, and Sturm came along our passage panting with the exertion of the ascent. As soon as he recovered breath, he said 'Vous avez un mémoire de Green; Monsieur Liouville me l'a dit.' So I handed it to him. He sat down and turned over the pages with avidity. He stopped at one place calling out, 'Ah voilà mon affaire'. So we turned over more pages and talked over the whole matter. Green's Essay made a great impression on Sturm and Liouville and others in Paris.
>
> When I got back to Cambridge I made enquiries and found all I could about Green's life, which is all I know up to this time. All that I could learn of his life is told in a little introduction by myself to Green's Essay, which I gave to Crelle for publication.
>
> You will find it in Crelle's Journal of about 1846.[17]

Enthusiasm in Paris, as Thomson noted in his letter, was not confined to Liouville and Sturm. The note in his mathematical diary for 27 January has the addition: '(Paris, Feb., Green's memoir creates a great sensation here. Chasles and Sturm find their own results and demonstrations in it.)' By the 1840s, the great generation of French men of science – so frequently referred to in this narrative – had mostly passed on. By 1845 Lacroix, Laplace, Fresnel, Fourier and Poisson were dead; of the remainder, Biot was seventy-one; Arago and Cauchy were still in their fifties. The newer generation, less brilliant, were consolidators, teachers and disseminators. Liouville was editor for forty years of the *Journal des Mathématiques Pures et Appliquées*, which he had founded in 1836; it later bore his name. August Crelle had founded his *Journal für die Reine und Angewandte Mathematik* in 1826, and edited fifty-two volumes before his death in 1855. William Thomson was familiar with both publications. He had spent his time between taking the Tripos examination and his departure for Paris in writing a paper for Liouville, which he had translated into French.[18] Crelle was so taken with the Essay that he offered at once to publish it in his

Journal.

Thomson knew nothing of George Green of Nottingham, but he lost no time in seeking information and wrote at once to Caius College from Paris, requesting information on Green. His query was passed to the Reverend J.J. Smith, who had been Assistant Tutor in 1833 on Green's arrival at Caius, and was now Fellow and Tutor. Smith had the College record on Green as undergraduate, graduate and Fellow, but may not have had much information concerning his pre-College life in Nottingham. He therefore wrote to Green's sponsor, Sir Edward Bromhead:

March 1845

My Dear Sir

You will have received the information that I am a candidate to succeed to the office of Librarian. The election will take place next month early: and I fain would hope you may be able to attend on that occasion, and that I may have your support. But, though I should have written this ere now – you will excuse this on account of the time – I am not even now writing on that ground: but I have just received through a friend a request from the Editor of the French Mathematical Journal to obtain information respecting [*illeg.*] first introduction to the Study of Mathematics, and his early progress in this pursuit. I recollect hearing that you were instrumental in bringing him [*illeg.*] College, if not the sole cause of his coming here: and I thought that you could therefore possibly, and if so that you would have satisfaction in giving the information in full to the requisitionist. Any information you may wish to have of the use to be made of what you supply, I will readily supply by inquiry.

Yours faithfully

J. Smith

Caius College
March 10

J.J. Smith is certainly writing in some confusion. The editor of the 'French Mathematical Journal' is, of course, William Thomson, writing from Paris; he had recently taken over editorship of the *Cambridge Mathematical Journal.* Smith's confusion is the more evident in that he omits to mention Green's name – an omission which Bromhead immediately queried, since a second letter followed, dated 17 March: 'I am sorry to have occasioned perplexity by my hasty omission. The subject of enquiry was G. Green, our late Fellow, whom I understand you were chiefly instrumental in bringing to the College.'[19] On reading the letters, a strong impression is one of surprise that the College apparently knew so little about one of its undoubtedly brilliant students only four years after his departure. Green, after all, had been expected to be Senior Wrangler, and thus bring an honour to a College which could claim all too few. Or does the limping syntax disguise a certain embarrassment, due either to the lack of knowledge the College could have been expected to provide, or to the fact that Green's last years in Caius had given cause for some anxiety? Another conclusion is that this is an indication of Green's reserved disposition, and he had never spoken of his pre-Cambridge life.

However that may be, Bromhead replied on 24 March, and Tomlin's equally valuable reply followed in April. It would appear that on his return from Paris Thomson was plunged into feverish activity: teaching, examining, corresponding with Liouville, writing articles, attending the meeting of the British Association, editing the *Cambridge Mathematical Journal*. In consequence, he had two letters from Crelle later in the year. Crelle wrote first in French from Berlin on 19 November 1845, thanking Thomson for the copy of the Essay and for copies of the *Cambridge Mathematical Journal*, and for the extract of the Essay drawn up by Arthur Cayley:[20] 'The work of Mr. Green will be most suitable for my Journal and will be published as soon as possible, and with the greatest care . . .' He followed this with a second letter on 27 December, written in German:[21]

> The paper by Mr Green, . . . is indeed very interesting and I will, as I have already promised, get it printed in my Journal; but I do have one small request. I think it might very well be appropriate to say in a few words in a short preface, why this paper should be reprinted after 17 years. The reason for this is the interest of its content, but it then should also refer to the fact, as I understand, that the paper had rather little circulation in public. I should like to write this preface myself of course, but it must necessarily be written in English; also in order to do so, I should have to study the work very thoroughly. Both of these aspects would provide great difficulty for me, the latter because I have too many tasks in hand and because of my poor health and advanced age, and

> the former because although I read English quite adequately, I cannot write or speak it well. Therefore I make so bold . . . to ask if you would add to your kindness by writing the preface which need only be quite short, and in which in addition to the analysis, which you have already made from the content, you include a few reasons for the reprinting and perhaps a few words about the deceased author himself and his scientific work.

Thomson, now a Fellow of St Peter's, replied on 11 February 1846:[22]

> Sir
>
> I received your letter of Dec 27th in Glasgow on the 8th January.
>
> I enclose a short introductory notice relative to Green's Essay on Electricity. To this I have annexed a chronological list of papers by various authors, since the time of its publication, on the same subject, or on subjects closely connected with that of Mr. Green's Memoir . . .
>
> I was quite unacquainted with the contents of Green's memoir and was quite unable to procure it at Cambridge till Jan 1845[23] when through the kindness of my former tutor, Mr. Hopkins, I obtained two copies. I was of course very much surprised to find that it contained the general theorem on Attraction besides other interesting matter and I have since endeavoured as much as possible to make the work known. Your kind cooperation will now insure its being generally known amongst scientific men . . .

There follows a short account – some fifteen lines – containing the general facts of Green's life as we know them, culled from College records and the letters from Bromhead and Tomlin.

Thus Green's Essay on Electricity and Magnetism found publication in Europe. It was too long to print in one issue of *Crelle's Journal*, so it was published in three parts: in volume 39 in 1850, volume 44 in 1852, and volume 47 in 1854.[24] It was later published in Berlin in a German translation, in Ostwald's *Klassiker der Exacten Wissenschaften* (1895). In England, the Essay had to wait for publication until 1871, when it was printed in *Mathematical Papers of George Green*, edited by Norman Ferrers and published by Gonville and Caius College.[25]

When Kelvin wrote his letter to Larmor in 1907, he added a further comment to his account of sending Green's Essay to Crelle: 'For many years I have wondered how I could have been so idiotic as not to publish it in the Cambridge and Dublin Journal.' The *Cambridge Mathematical Journal* was founded in 1836;[26] unlike the *Transactions of the Cambridge Philosophical Society*, it published articles by undergraduates as well as graduates. Thomson had contributed papers as a student under a pseudonym, and on his graduation in 1845 he took over the editorship from Ellis, at a time when discussions were under way for amalgamation with the *Dublin Mathematical Journal.* The publication of Green's Essay in a German journal would have very limited circulation in England, and despite Thomson's zeal in making Green's work known, his reputation would have been enhanced if the Essay had been as widely available in scientific circles as his later memoirs. Electricity and magnetism, still neglected subjects in Cambridge, were about to become of major importance and general interest with the work of Thomson, Faraday and Clerk Maxwell. It is indeed regrettable that Green's Essay was not available during this important quarter of a century.

William Thomson did not stay long at St Peter's. In July 1846, applications were invited for the Chair of Natural Philosophy in the University of Glasgow. Thomson submitted an impressive number of testimonials, written by a number of impressive people, some of whom he had recently met in France; they included the no doubt significant one from Regnault. There was also one of particular significance from Liouville who, after referring to Thomson's abilities and researches, states that he considers him 'worthy to replace the eminent geometer, so little known in his lifetime and so worthy of a better fate, the illustrious George Green'.[27] And in a private letter of support to Thomson at the same time, Liouville wrote as follows:

> By the nature of your talent, you seem to me called upon to repair the loss science has received in the person of your fellow countryman the illustrious Green, whose merit you have been one of the first to recognize and to whom I myself endeavour on every occasion to give the recognition he was denied in his lifetime. You will be more fortunate than he . . .[28]

Thomson, from the first, worked to make Green's name known.

Even in the first paper he had taken to Paris for Liouville's *Journal*, he had immediately included a reference to Green's Theorem as given in the Essay. On his return to Cambridge, Thomson read a paper on the theory of magnetism to the British Association – meeting in Cambridge in June 1846 – in which he included a reference to two English men of science whom he was to revere throughout his life: 'If the laws of Coulomb are assumed, then by very simple analysis first given by Green, we arrive at the laws of Faraday as theorems.'[29] A few years later (in 1849) Thomson, not yet a member, submitted to the Royal Society another important paper on the theory of magnetism; this followed detailed correspondence with Faraday, whom he had met at the British Association meeting in Cambridge some years earlier.[30] William Thomson was appointed to the Chair of Natural Philosophy at Glasgow at the age of twenty-two; he occupied it for fifty-three years. To the end of a long life he repeatedly expressed his admiration for Green, regarding him, Fourier and Faraday as his three great heroes of science.

Thomson himself edited the *Reprint of Papers on Electrostatics and Magnetism* (1872), nearly six hundred pages of collected papers written by him and printed during the previous thirty years in various journals and Transactions: *The Cambridge Journal* and *The Cambridge and Dublin Journal*, *Liouville's Journal des Mathématiques*, the *Philosophical Magazine*, Nichol's *Cyclopaedia*, the reports of the British Association, the *Transactions* of the Royal Societies of London and Edinburgh, and the Philosophical Societies of Manchester and Glasgow. This collection is of special interest, since Thomson's editorship meant that he commented – in voluminous footnotes, and considerably later – on his ideas and findings of earlier years. References in the text reveal Thomson's continued regard for Green, and his debt to him: 'The mathematical theory [of electricity] received by far the most complete development which it has hitherto obtained, in Green's Essay . . .'[31] 'According to Green's remarkable theorems, triply re-discovered by Gauss, Chasles and the writer of this article . . .'[32] 'This result was first given by Green, near the conclusion of his paper, "On the Laws of the Equilibrium of Fluids" . . .'[33]

> Green's Essay on Electricity and Magnetism and his other papers on allied subjects contain, beside the solution of several

> problems of interest, most valuable discoveries with reference to the general Theory of Attraction and open the way to much more extended investigations in the Theory of Electricity than any that have yet been published.[34]

Thomson's friend George Gabriel Stokes followed his tutor William Hopkins's advice on graduation in 1841 – to pursue the study of hydrodynamics – and did little on electricity and magnetism, considering them the domain of his friend Thomson. Stokes was appointed to the Lucasian Chair of Mathematics in 1849. As professor:

> Stokes was a pivotal figure in furthering the dissemination of French mathematical physics in Cambridge. With Green, who in turn had influenced him, Stokes followed the work of Lagrange, Laplace, Fourier, Poisson and Cauchy. This is seen most clearly in his theoretical studies in optics and hydrodynamics'.[35]

Stokes, like Thomson, makes important references to Green – in this case to his seven papers on the behaviour of waves and the reflection and refraction of light, which would reveal important links with each other. Such references – like Thomson's – repeatedly establish Green's priority as the first in the field: 'The nature of the motion of the individual particles . . . was first taken notice of, I believe for the first time by Mr. Green . . .'[36] 'Moreover, the laws of the motion of a solitary wave, deduced by Mr. Green for the theory of long waves . . .'[37] 'In the 6th volume of the Transactions of the Cambridge Philosophical Society, p. 403, will be found a memoir on the reflection and refraction of sound, which is well worthy of attention . . .'[38]

Stokes's second Report to the British Association, given in 1862, dealt with his other main interest, optics; it was entitled 'Report on Double Refraction'. Here again there are frequent references to Green's later papers on the reflection and refraction of light:

> I come now to Mr. Green's theory, contained in a very remarkable memoir 'On the Propagation of Light in Crystallized Media' read before the Cambridge Philosophical Society on May 20th 1839, and accordingly by a curious coincidence, on the very day that Cauchy's second theory was presented to the French Academy. Besides the great interest of the memoir in relation to the theory of light . . .[39]

> In this beautiful theory therefore we are presented with no forced relations like Cauchy's relations; the result follows from the hypothesis of strictly transversal relations, to which Fresnel was led by physical relations.[40]

Stokes had paid his most comprehensive tribute to the corpus of Green's work however, in his first Report on Hydrodynamics:

> Indeed Mr. Green's memoirs are very remarkable both for their elegance and rigour of the analysis, and the ease with which he arrives at most important results. This arises in great measure from his divesting the problems he considers of all unnecessary generality: where generality is really of importance, he does not shrink from it.[41]

In the light of these tributes from Thomson and Stokes, Whittaker discerns Green's far-reaching influence on the generation which followed him:

> It is impossible to avoid noticing throughout all Kelvin's work evidences of the deep impression which was made on him by the writings of Green. The same may be said of Kelvin's friend and contemporary, Stokes, and indeed it is no exaggeration to describe Green as the founder of that 'Cambridge School' of natural philosophers of which Kelvin, Stokes, Rayleigh, Clerk Maxwell, Lamb, J.J. Thomson, Larmor and Love were the most illustrious members in the latter half of the nineteenth century.[42]

When Rutherford first observed that one element could be artificially transmuted into another in 1919, and thereby created a new branch of science in nuclear physics, Green's work would appear to have passed into science history, along with that of several other nineteenth-century men of science. The advent of Einstein and the new discipline of relativity and quantum mechanics appeared to confirm this, until there was a second rediscovery of Green's work by a group of scientists – American, British and Japanese – who were investigating the interactions between particles, using quantum mechanics.

In the late 1940s and early 1950s several articles appeared in the *Physical Review* and other journals, by J. Schwinger, R.P. Feynman, F.J. Dyson, S. Tomonaga, T. Matsubara and others, in which forms of Green's functions were used to great effect.[43] They soon proved to be an efficient and powerful tool in solving problems encountered in the application of quantum mechanics to solid state physics as well as nuclear physics. This led to their widespread use – which continues today – in the detailed understanding and exploitation of the 'high-technology solids', such as semiconductors and superconductors.[44] In classical physics Green's functions had already provided a most efficient way of describing how a system responds to a force, so it was perhaps inevitable that they would eventually be used to describe the behaviour of atomic and nuclear forces.[45]

The impact of Green's work is not restricted to Green's functions. His work in electricity had been developed by Thomson and Clerk Maxwell who, with Faraday, laid the foundations of modern electromagnetism. More recently, his work on light and sound has contributed ultimately to the use of ultrasonic diagnostic techniques in medicine and in seismic soundings in geology, and, through his theory of total internal reflection, the use of optical fibres in telecommunications.[46] The ideas he contributed to elasticity[47] underpin the theories behind much of today's civil, mechanical and structural engineering, as well as the detailed properties of solids.

It might, therefore, be no exaggeration to say that to practising scientists, Green is now something of a giant. His achievements, as exploited in the twentieth century, are not as well known to the layman as those of Kelvin and Faraday, but though they are fewer in number, they yield nothing in their intellectual profundity to the genius of Stokes or Clerk Maxwell. Green, who has the almost unrecognized distinction of having corrected a mistake in Newton's calculations,[48] has remained in the shadows as far as popular acclaim is concerned. A passage from Whittaker was quoted above in which the names of Newton and Green were polarized, but George Green's fate has not been as fortunate as Isaac Newton's. Newton enjoyed the fruits of fame during a long and honoured lifetime, and a posthumous reputation of Handelian proportions. George Green, on his death, earned a modest paragraph in a local newspaper as his sole obituary, and a grave neglected and forgotten.

For men such as William Thomson and John Frederick Herschel, entry into a distinguished public and scientific life was easy, since they came from successful and well-established academic families. From their youth they were surrounded by books and learning, and given every encouragement and facility to develop their abilities. Charles Babbage, precocious in his youth, was aided in every way by a well-to-do banker father. The fact that George Green was denied these advantages and was largely – or entirely – self-taught makes the value and extent of his achievements all the more impressive, but his case is not unique.

George Boole came from humbler circumstances than Green, and was also largely self-taught, though it is probable that he, like Green, was encouraged by Bromhead. Boole, however, was more fortunate than Green in his lifetime, achieving a university appointment, public recognition, and a range of publications. He, like Green, left behind an eponymous mathematical tool, boolean algebra.

Michael Faraday is another example of an outstanding figure emerging from a modest background. He too was self-taught, but he had the advantage of early adoption by Sir Humphry Davy as his laboratory assistant at the Royal Institution, from which he went on to establish a wide and public reputation.[49]

It is well worth remembering that in his forty seven years of life, Green had been able to devote fewer than ten to full-time study, and of those, three were taken up with the acquisition of his degree. That he managed to include so much of value to mathematics and science in only ten papers, published in the space of eleven years, is remarkable.

Chapter 11

'Honour in His Own Country'

For over a hundred and twenty years Green's Mill had stood, a gaunt and derelict landmark, on Belvoir Hill, with Nottingham to the west, Colwick Hills to the east, and the River Trent to the south. Built originally near the village of Old Sneinton, by the early 1820s it was partially surrounded by the houses of New Sneinton, which had started to cluster round St Stephen's Church. After the Enclosure Act of 1845, when Sneinton was incorporated into the town of Nottingham, its population grew, though it still retained some of its farms and remained a desirable residential area. After the First World War, however, its character changed, and by the 1950s Sneinton had deteriorated into one of the poorer problem areas of the city.

It is not surprising, therefore, that Clara Green's mill attracted no attention and became a symbol, in its neglect, of the ignorance and indifference the town accorded the man who, for most of his life, had worked it. A photograph from the 1860s shows the sails still in place, but its decline was thereafter recorded by local artists. T.W. Hammond drew it in 1901 – in what one suspects is rather a romanticised setting – when the wooden cap was rotting, obviously a haven for the birds, but the fantail was still in place. Sixteen years later, a local schoolmistress exhibited a pastel drawing in the Nottingham Artists' annual display, but by then the fantail had disappeared. Karl Wood, a regional artist, who aimed to paint all the windmills in the Midlands, produced a watercolour in 1937. This showed the mill as repaired by Oliver Hind, a Nottingham businessman and philanthropist, who had purchased it from the Crown in 1920. The cap was covered with copper, and the mill was let out as a furniture-polish factory. At the same time, a plaque was placed on the wall with the help of the Holbrook Bequest, which read – not quite accurately –

HERE LIVED AND LABOURED
GEORGE GREEN
MATHEMATICIAN
B. 1793–D. 1841

The final line of the inscription, as originally composed, was, in the event, omitted: 'Here is honour in his own country'. This was perhaps just as well, given the subsequent neglect of the mill. Not surprisingly, in view of the nature of its contents and its singular shape, the inevitable finally happened. In 1947 it caught fire and was completely gutted. Oliver Hind, with great foresight, sealed the shell by boarding up the windows and doors and placing a flat concrete top where the cap had been. And there the mill stood for the next thirty years.

In 1924 there was an early attempt to associate George Green with his mill. A local writer, S.F. Wilson, published an article on Green in *The Miller*[1] with two photographs of the mill – one from about 1860, with the sails and gallery still in place; the other showing the workmen covering the dome with copper. Wilson had measured the mill, and he gave a detailed description of its state at this date before embarking on an account of Green's life. This was based on Dr. Becket's paper and local newspaper reports, but it is of interest on several counts.

Wilson postulated that Green's studies in mathematics were influenced by his observation of the mechanical processes of milling. He substantiated this claim by launching into an arithmetical calculation of the force required to hoist a sack of grain from the ground to the top of the mill – a calculation which, he felt, would lead Green to the idea of the conservation of energy, first introduced in the Essay of 1828. Wilson – like Professor L.J. Challis of Nottingham University and his colleagues fifty years later – sought to establish Green's reputation in his native town, and proposed the institution of 'Memorials of Green'. There should be, he suggested, a Green's scholarship at the proposed 'East Midlands University',[2] and an annual Green's Prize in mathematics. Furthermore, he advocated a concerted move to secure an endowment for a permanent scholarship from the money, or its equivalent, which had accrued to the Crown when it took over the mill and its surroundings after Clara's death, which had taken place only five years before.

Wilson was not alone in his advocacy of Green. In University College Nottingham Dr Edith Mary Becket was still active, with the interest and support of Professor H. Piaggio, Professor of Mathematics, and Professor F. Granger. Dr Becket was a member of the Nottingham Subscription Library, and it is thought that she was the one who discovered Green's association with Bromley House. In the 1920s the Council of Bromley House Library, of which Dr Becket was a member, disposed of a large number of old books. The Minute of 22 November 1927 empowered the Secretary, W. Heazell, 'to get rid of them to the best advantage he was able'. In 1930 a collection of 'books and documents' was passed over by Mr Heazell to the University College Library – unfortunately before the present accessions record of the Library was established.[3] It is conceivable that Dr Becket promoted this transfer, but it is unlikely that the collection contained any material pertinent to Green, as she would presumably already have identified it in her memoir. Certainly no books of Green's period are now housed in the Rare Book Room of the University Science Library – except the Royal Society *Transactions*, which may have been acquired from another source.

Professor Granger was now Vice-Principal of University College, but he maintained his interest in Green. In his *Memorials of the University College, Nottingham*,[4] he wrote:

> The most important intellectual event in the history of Nottingham was the publication by George Green in 1828 of his Essay on the Application of Mathematical Analysis to the Theories of Electricity and Magnetism. After writing his masterpiece, Green, the most important English mathematician since Newton, went to Cambridge. The present year will mark the centenary of the Essay. It would be an act of justice to give his name to a Chair of Mathematics. At this anniversary of George Green's first publication, Albert Einstein sends expressly his hearty good wishes for the success of the University College.

Einstein's contact with the University College was apparently Dr H.L. Brose, a Rhodes scholar from South Africa, and Reader – later Professor – of Physics. It was perhaps through him that Einstein was moved to send a telegram of congratulation to the City marking the centenary of the Essay's publication. This was noted in a few modest paragraphs in the local newspapers; otherwise the event appears to have gone unnoticed.

Two years later, Einstein arrived in person in Nottingham to deliver a lecture at the University College, and Professor Granger was his host. Professor Brose translated Einstein's lecture on relativity from the German to the English audience. The chalk and blackboard still bearing Einstein's calculations are now treasured possessions of the University Physics Department, which celebrated the fiftieth anniversary of Einstein's visit in 1980. Einstein knew and appreciated the value of Green's work. As Professor Granger wrote in a letter to the *Nottingham Journal* of 30 June 1930:

> I showed Professor Einstein my copy of Green's Essay. He turned to the famous paragraph on page one[5] and remarked on the way Green had anticipated the work of later mathematicians, especially Gauss.

In parallel with the local story of Green using the top floor of the mill as a study is a more recent piece of folklore: that Einstein made a pilgrimage to Green's grave in Sneinton. It would be gratifying to find that he had. These, however, are the facts: Einstein had said that he would arrive in Nottingham at 3.45 p.m. on the day of the lecture, which was timed for 7.00 p.m. It had been arranged that he would be welcomed with tea, then taken to Sneinton to visit Green's grave. In the event Einstein did not arrive until 6.30 p.m., having stopped at Grantham, according to one report, to visit Newton's birthplace at Woolsthorpe.[6]

In 1937 the British Association for the Advancement of Science held its Annual Meeting in Nottingham. H.G. Green gave an address on George Green to the Mathematical Section, followed by a visit to the mathematician's grave in St Stephen's churchyard, Sneinton. The members of the conference who made the pilgrimage were moved to protest at its condition to the Lord Mayor of the City. In consequence the stone slab was repaired, and the inscription was reincised and inlaid for permanent legibility; it has since been maintained in good condition by the City authorities. Later the grave, and lettering on the adjacent grave housing the remains of Jane Green, were cleaned and recarved in time for the official opening of Green's Mill and Science Centre in 1985. These graves under the horse chestnut trees in the north-east corner of St Stephen's churchyard now offer a quiet oasis amid the busyness of modern Sneinton, and an invitation for quiet reflection to those pondering the fortunes of George Green and his family.

Sporadic interest had been shown by Cambridge in the person of Sir Joseph Larmor who, as early as 1907, had written to Lord Kelvin. This letter elicited the one (quoted above) in which Kelvin describes his discovery of Green's Essay. Larmor was also in touch with Professor Granger, now Principal of the University College. Clara Green's death in 1919 revived the enquiry into Green's Nottingham connections and in particular the problem of reconciling his Fellowship at Caius with the presence of his family in Nottingham and his son George Green's degree at St John's College. The quest led both to Dr Becket's correspondence with Dr Scott, President of St John's, and to a renewal of the correspondence between Larmor and Granger, in which the Reverend F.C. Finch, vicar of St Alban's, Sneinton, also joined. His interest in Green, as he wrote to Larmor, was the fact that his uncle, Edward Brumell, had been Third Wrangler to Green's Fourth, and was later President of St John's.

Finally, in 1928, Cambridge was again in contact with Nottingham. On this occasion Professor F. Stratton, of Gonville and Caius College, wrote to H.G. Green, Reader in Geometry at Nottingham University, about a proposal to publish a *de luxe* edition of Green's *Mathematical Papers* to replace Norman Ferrers's 1871 edition, which was now out of print. Larmor, who was close to retirement, had offered to produce historical annotations to the *Papers*. He was willing for certain letters to be used, wrote Stratton, but firm in counselling discretion on the subject of Green's private life.[7]

H.G. Green was invited to contribute a suitable biography of Green, since he had collected so much information about his Nottingham background. This was a most painstaking piece of research, as the Appendices to his published account reveal. Unfortunately, news came from Paris that a similar edition was envisaged, and the Cambridge proposal fell through. The French venture, if it ever started, was apparently abandoned on the outbreak of war in 1939. In the early 1940s, however, H.G. Green was approached for a contribution on George Green by the compilers of a symposium to be published in New York in 1945 and presented to Professor George Sarton of Yale University, a noted historian of science, on the occasion of his sixtieth birthday. H.G. Green collated the results of many years' work into a short biography and a number of appendices listing details of Green's work, family dates, movements and possessions, together with the letters from Tomlin and Bromhead and the correspondence between Kelvin, Crelle and Larmor. Unfortunately, this volume has been out of print for many years, so H.G. Green's contribution has had only limited circulation.

So for the thirty years that the mill still stood neglected in Sneinton, there was also a dearth of information on Green for any occasional enquirer, both mill and mathematician suffering a like fate. Then in the 1970s came the ambitious project – again emanating from the University of Nottingham – to rehabilitate both in the one venture. Two quite separate events in Nottingham started the process. In 1974 a Physics Conference on magnetic resonance, named after the French physicist André Ampère, was to be held in the City, and the Physics Department of the University organized, in co-operation with the City Leisure Services Department and the Keeper of Art at the Castle Museum, the first exhibition ever on the work of George Green, who had, like his contemporary Ampère, worked on the properties of light.

For Professor L.J. Challis and his colleagues in the University Physics Department, this presentation of Green's achievements sprouted two important offshoots. In the same year (1974) Professor Challis attended an International Physics Conference in Budapest and addressed the Plenary Session on George Green, showing slides of the exhibition and the ruined mill. As a result, the Chairman of the Conference sent a telegram to the Lord Mayor of Nottingham, expressing their appreciation of the City's commemoration of Green. This was the second attempt by scientists in nearly forty years, following the protest of the Mathematical Association in 1937, to arouse the City's awareness of Green's national and international importance. The second development from the Castle Museum Exhibition was the publication in 1976 of the booklet *George Green, Miller, Snienton*. This was in three sections: a short account of the scientific significance of Green's work by Professor L.J. Challis and his colleagues Dr R.M. Bowley and Dr F.W. Sheard; detailed information on the Green family compiled by the Senior Archivist in the City, Mrs F. Wilkins-Jones; and a review of Green's academic career by David Phillips, Keeper of Art at the Castle Museum. It covered much of the ground previously researched by H.G. Green, but it was especially valuable for the inclusion of recently discovered letters from Green to Bromhead transcribed by David Phillips, who was also the editor.

The second event in Nottingham which furthered the 'Green Movement' also occurred in 1974. The City published plans for a revised town transport system, according to which a proposed bypass through Sneinton threatened the 'Green area'. The plans were later abandoned, but not before rumours circulated that Mill House would be demolished, and possibly also the mill itself. These prompted Professor Challis and his associates to protest at the proposed demolition and enclose a report from the late Dr Norman Summers of the University Department of Architecture on the viability of the mill's restoration. On receiving the report, the Nottingham City Planning Department invited co-operation from local voluntary organizations in reviewing the situation. As a result, Professor Challis formed the George Green Memorial Fund, which – co-operating with the City Planning Department and Leisure Services Committee, and with the Sneinton Environmental Society and the Nottingham Civic Society – went to work to raise funds for the restoration of the mill and the establishment of a permanent museum showing the life and work of George Green.

The campaign was to be a long one, taking over a decade, but by 1977 the strategy was taking shape. In that year the Corporation of Nottingham undertook a comprehensive survey of the Sneinton area, with its run-down property and general air of neglect. A high priority for this densely populated neighbourhood was a park and a children's playground, and the old Green's Gardens on the slopes surrounding the mill seemed ideally suited for this. The restoration of the mill also attracted great local support and interest, not least among the schools in the area. The City therefore welcomed the initiative of the George Green Memorial Fund, and agreed to provide the cost of the park and, in principle, that of the restoration of the mill. The mill, however, was still the property of the Oliver Hind Trust, from which the Fund eventually bought it. In November 1979 the mill and mill-yard were formally handed over to the Lord Mayor by Professor Challis, and became the property of the City of Nottingham.

The restoration started with clearing the mill and yard of thirty years' debris and undergrowth. The brick walls of the tower were repointed, doors and windows were replaced, and a new cap was constructed in the yard. On a freezing day in December 1981, this was hoisted by power crane to the top of the fifty-foot tower – three hours behind schedule, since the diesel oil in the crane-lorry kept freezing up and had to be thawed out with the help of a one-bar radiator, borrowed from the workmen's hut! Meanwhile, the interior machinery was being constructed, and millstones were brought over from a disused mill in Lincolnshire. The restoration, originally in the hands of an independent millwright, was taken over at this point by Messrs Thompsons of Alford, Lincolnshire, and a period of two years followed before the sails, the most evocative and spectacular part of the mill, were finally put in place. Thompsons also provided more sophisticated machinery, since by now the project had fully taken flight.

Green's Mill was not to be restored solely for topographical or sentimental reasons, with occasional bursts of activity on fine weekends with the help of a few volunteers, which was all the early Green enthusiasts had thought possible. It was now to be a fully working mill, grinding daily; this required higher specifications to comply with Health and Safety Regulations, and therefore a more thorough and long-term restoration than had originally been envisaged. Furthermore, the Green Museum, originally conceived as occupying the ground floor of the mill, was to be housed separately in buildings sympathetically designed and constructed on the site of the old mill's outbuildings. There would be exhibits on mills and milling, a seminar or classroom for teaching projects and school visits, and – most importantly – the George Green Science Centre, an area of push-button, hands-on working models, where visitors, particularly youthful ones, could make electricity, play with magnets, and experiment with apparently simple apparatus – in one case working, did they but know it, on the principle of total internal reflection.

In September 1985 Professor Challis addressed a similar gathering of world scientists to that of 1974, again held in Budapest. On this occasion he was able, eleven years after his previous visit, to show slides of the restored mill and the Science Centre. On this occasion, too, the Conference unanimously sent a telegram to the Lord Mayor of Nottingham, congratulating the City on the completion of the enterprise. The object of the George Green Memorial Fund had been achieved. 'What we are striving for', wrote Professor Challis, a few years ago, 'is a living memorial to Green which will both help people to appreciate the remarkable achievements of a man who was very largely self-taught and also stimulate a lasting interest of young people in physics and mathematics.'

The restoration of the mill and the establishment of the George Green Science Centre was a triumphant testimony to successful co-operation between academic interest, municipal enterprise and voluntary effort. Many science organizations, both national and international, contributed some £40,000 to funds for the museum. The Nottingham City Planning Department worked with commitment and enthusiasm, and co-operated readily with the George Green Memorial Fund and the Sneinton Environmental Society. The plans for the restoration of the mill complex, submitted under the aegis of the Nottingham Civic Society, gained a Heritage Award. Many business concerns, both local and national, helped in kind – whether with the mill, the museum buildings or the science models.

One of the more unusual – but vital – offers received was for lightning conductors – one for the central finial on the top of the cap; one for the fantail; and one for the tip of each of the four sails. This was an anachronism, certainly, as are some of the provisions necessary to comply with the Health and Safety Regulations, and a measure unknown to Green Senior as well as his son.[8] One of the most deeply appreciated was the gift of fifty copies of a facsimile edition of Green's Essay. These came from Professor Ekelöf of Sweden, who in 1958 had commissioned a limited edition of a thousand copies of the Essay published in Nottingham in 1828. A further consignment was later acquired by Professor Challis, and formed a valuable and lucrative addition to the assets of the George Green Fund. To date these are the only facsimile copies of the Essay available.

The complex was officially opened on 6 July 1985 in the presence of the Lord Mayor and the High Sheriff of Nottingham. Professor Challis, most appropriately, was asked to open the mill, since it was largely his vision and enthusiasm, supported by his university colleagues, which had initially launched the project of restoring the mill as a memorial to Green. A personal and family dimension to the event was added by the presence of Mrs Eva Saunders, the mathematician's great-granddaughter, and members of her family. Equally appropriately, Professor Sir Sam Edwards was invited to open the Green Science Centre. Sir Sam was the ideal person to approach for such an occasion, since in his Cambridge appointments he represented much of what George Green had laboured and stood for: Director of the Cavendish Laboratory, named after Henry Cavendish, from whose paper of 1771 Green quoted in the opening paragraph of his Essay; University Professor of Theoretical Physics, and finally Fellow of Gonville and Caius College. In paying tribute to Green's contribution to science, Sir Sam cited three leading nineteenth-century scientists, making the point that Faraday's experimental work and Clerk Maxwell's theoretical work depended for their fruition on the mathematical framework provided by Green.

Green's Mill and Science Centre is now one of the most popular of Nottingham's many attractions, visited by tens of thousands of people a year. It is a notable example of an industrial archaeological restoration, an education project, and a tourist attraction. The sight of mill sails turning against the sky, viewed while driving into the city, is quite hallucinatory, and seen at night from surrounding villages, illuminated on the city skyline, it is magical. The genteel neighbourhood of Green's Gardens and the solid respectability of Belvoir Terrace, where lived the Tomlins and their like, have long since vanished, but something of the ambiance of leisure and relaxation has been restored, and Sneinton residents, who once gave their address as Nottingham, are now proud to say that they live in Sneinton and welcome the many visitors to their mill.

Visitors now number nearly 45,000 per year and the figure continues to increase. The Mill is particularly popular at weekends and holiday times when families of children arrive to 'make electricity' in the George Green Science Centre and run their fingers over the plasma ball. Adult groups from local history societies book a special visit for a summer's evening excursion. Foreign students and University guests—from Holland, Scandinavia, Russia, the U.S. and Asia—have come to pay their respects to Green in company with their British colleagues.

The most numerous, however, are from the local schools and colleges—some 140 a year, ranging from infant classes to students in higher education. In an era of conservation and in an urban setting, the Mill has a strong educational function, with obviously the more accessible aspects of mills and milling attracting the younger children, while there are more scientific challenges in sessions designed for high school students. Thus the Museum's Education Officers participate in the annual National Science Week and arrange talks by, for example, Professor Lawrie Challis on 'Physics at Low Temperatures' or his colleague Peter Coles, Professor of Astronomy, who in 1999 gave a topical lecture on 'Einstein and the Eclipse'. A pilot project based in a local technological high school, supported by both Nottingham and Nottingham Trent Universities, resulted in students constructing a model wind generator and a power pump for the school pond, and this collaboration is being developed.

The Mill grinds twenty tonnes of flour a year and stone-ground flour is on sale at the shop, as well as a book of recipes, since monthly baking sessions are organised for children and adults. Other events are regularly organised for weekends and vacations. Special favourites are the Harvest celebrations and Christmas Carols at dusk in the mill-yard by the light of the illuminated Mill. On one occasion an evening Nature Watch resulted in the identification of a rare night moth: to the east the Trent valley can be seen from the Mill gallery and to the north Colwick Woods are barely a mile away on an adjacent hill. On another evening, radio hams set up their equipment in the Mill, taking advantage of its height and isolation to contact fellow enthusiasts as far away as Eastern Europe. The latter may have found various vantage points for their broadcasts, but few are likely to have emulated those in Sneinton who had the advantage of being able to attach their aerial to the tip of a mill sail sixty feet from the ground! After nearly fifteen years of continuous use however, the Mill has needed repairs. Its exposed position and the pollution of an urban environment have had their effect. Mill construction, if not entirely a lost art, suffers from a lack of reliable building materials common in Green's time; thoroughly seasoned wood for example, and single poles long enough to provide the thirty-foot long sail stocks, were particularly hard to come by. The inside of the Mill had to be repainted, since the whitewash used was found to contain a lead ingredient unacceptable to the Health and Safety Inspectors. The City Council has now started a programme of regular maintenance under the expert advice of the miller, who shares a family enthusiasm for mill restoration and is an excellent and well-informed guide for visitors of all ages.

The aim of those restoring the Mill was to re-establish it as a memorial to George Green and draw attention to his life and scientific importance. Although the Mill was restored to reproduce the original as faithfully as possible (no detailed records existed), it was never envisaged to simulate the bustle of the Mill in Green's time, with the horses and carts, the stables and harness rooms, the grain-stores, the mill-hands and the manager's office. The hard round cobbles—still there in the 1970s—were replaced by paving stones, the former being judged too hazardous for children, the elderly and disabled visitors. Instead the curious attraction of Green and his Mill has led the mill staff, who welcome the visitors and care for the Mill, to transform the mill-yard into a quiet restful area, with small beds of English herbs and plants which may well have grown in the garden of the Green family house opposite. Such plants are for sale and these, with perhaps the purchase of a cup of coffee, contribute to their providing further amenities. They have even launched a local group of Friends of Green's Mill.

The Green family house has had a more controversial history. Apart from a period when Clara Green, the owner, is reputed to have lived there as a paying guest of her tenant, it was never inhabited by a member of the Green family after George Green's departure for Cambridge in 1833. On Clara's death in 1919, the family house was sold. In recent times the owner was Mr. Harold Wiseman, a Green enthusiast, who placed a commemorative tablet on the wall of his house, known as Belvoir Mount in Clara's time, but which he renamed Mill House. His nephew, who lived with him, inherited the house and later offered it for sale to the City Council.

This offered considerable possibilities—of rounding off a historic complex, establishing a detailed display of life in early nineteenth century Nottingham and in particular of the baking and milling trades, even much needed storage space for the Mill. Unfortunately, the funds for purchase and maintenance were not forthcoming and the offer had to be turned down. Shortly afterwards the owner died, leaving the house to a close friend. Once again, in a little over two years, Mill House was on the market. Professor Challis endeavoured to interest the National Trust but inspection showed that the interior had been too much altered to consider any possibility of historical restoration.

When seen some ten years ago, the interior was evocative of a prosperous artisan's house of the early 1800s. The rear door from the mill yard led into a stone-flagged scullery with a large stone sink and a sunken floor; across the passage was a large, rather stark kitchen with a small office or counting house built on at the back. The two large front rooms either side of the front door would have served for family gatherings and provided an atmosphere of prosperous respectability—characteristic of the rising artisan class of the time. But *autre temps, autre moeurs.* Early nineteenth century domestic conditions do not meet the requirements of late twentieth century living. Mill House is in sound condition and well cared for. The relationship between House and Mill is on a good footing, which is helpful when each share a narrow road winding up Belvoir Hill from the main thoroughfare below. On the opposite side of the road, on the other side of St. Stephen's church wall, lie the Green family graves. In 1993 the City Council cleaned George Green's grave in preparation for the bicentenary celebrations. This was the first time for thirty years, and the inscription inlaid with a black cement was then still visible. By 1999 however its condition gave cause for concern. The felling of several of the splendid horse chestnut trees round the graves had left them unduly exposed to the weather; indeed, the inscription on Jane Green's grave alongside was quite illegible. It was not possible, for technical reasons, for the letters to be re-carved on the original stone, so the George Green Memorial Fund arranged for the inscription to be cut on a smaller stone fixed to the lower part of the grave. (In 1993 they had Clara Green's grave in the Rock Cemetery repaired). The City Council, with contributions from Mrs Elizabeth Draper (one of Green's descendents), the Sneinton Environmental Society and the Friends of Green's Mill, undertook the restoration of George Green's grave, and the Sneinton Environmental Society also planted trees near to the graves.

A highly significant recognition of George Green's life in Nottingham was the naming of the George Green Room in the Nottingham Subscription Library in Bromley House. The room had housed the library of the Nottingham Law Society for many years and on its removal the Council of the Bromley House Library named the room in honour of its most illustrious subscriber. This was a notable gesture, since the Library itself is now one of the most interesting features of Nottingham. It celebrated its 175th anniversary in 1991 with a slim volume of essays on this historic institution—it is one of only fifteen Private Subscription Libraries still in existence.[10] The George Green Room contains an enlarged photograph of Green's Mill painted earlier this century by a local artist, occupying the space between two windows which would have been occupied by a portrait of Green, had there been one. There are also photographs of the restored Mill and of Jane Smith, the mathematician's eldest daughter, and a collection of photographs and documents relating to Green's life and subsequent events.

In 1988 it was realised that 1991 was the 150th anniversary of Green's death, but also that 1993 would offer a more prestigious celebration of the bicentenary of his birth. Accordingly, the earlier anniversary was marked by a more modest ceremony, for which members of the George Green Committee and the Sneinton Environmental Society gathered on the morning of 31st May at Green's grave in St. Stephen's churchyard in Sneinton. The Vicar read a prayer and Professor Challis spoke a few words commemorating George Green and his work. The occasion raised an awareness of the singularity of Green's life—a working miller, now an internationally recognised scientist, who had spent so many years of his life in the village of Sneinton. Green's Mill and the family house were only a few hundred yards away, and Jane Smith's house, in which he died, was just across the road. The pall bearers would have carried the coffin from house to grave—and even in Cambridge, Green ('son of a miller') had taken his trade with him.

This humble ceremony in Sneinton contrasted sharply with the local and national—and even international—nature of the celebrations for the bicentenary of George Green's birth. Professor Challis, in collaboration with influential scientific colleagues, was soon planning events for 1993, and long-term plans resulted in their taking place in Nottingham, Cambridge and London. The Bicentenary Celebrations opened on Tuesday evening, 13 July 1993, with a Civic Service of Thanksgiving in St. Stephen's church. The Lord Mayor and the Sheriff of Nottingham represented the City; the Vice-Chancellor, Nottingham University, and Professor Julian Schwinger and Professor Freeman Dyson represented the international community of scientists. From Cambridge came Professor Sir Sam Edwards, F.R.S., President of Gonville and Caius College, and Revd Dr John Polkinghorne, F.R.S., President of Queens' College.[11] Dr Polkinghorne, a highly respected mathematician and theologian, gave the Address. This followed a brief but cogent Tribute from Professor Challis who, in layman's terms, highlighted some of Green's most important achievements. A Plaque in the wall above the Pulpit was then dedicated by the Assistant Bishop of Southwell;[12] it bears the inscription: In Commemoration of the Bicentenary of the Birth of George Green, Miller, Mathematician and Physicist 1793–1841. The reading was given by Mr Tom Huggon, Chairman of the Sneinton Environmental Society, and was taken from the little known Chapter 39 of Ecclesiasticus which seemed, in biblical terms, to suggest something of what George Green had achieved and what he now represented.[13]

What was memorable on this occasion was the meeting for the first time of the two branches of the Green descendants, who had been unaware of each other's existence until eight years previously: two members had come from Canada and one from New York—in all, twelve blood descendants who with their families numbered twenty-five people.[14] After the Service, flowers were laid on Green's grave by the family, and the civic and academic dignitaries attending the service. The loyalty and steadfastness of the faithful Jane Smith was also commemorated with flowers; she lies in a grave next to the mathematician's with her daughter Catherine and son George who were buried before her. Later some of those present went up to the Mill, which was splendidly illuminated as usual against the night sky and specially open for the occasion.

Next morning, the 14th of July and Green's actual birth date, there were festivities at the Mill, now decorated with flags from sail-tip to sail-tip. There was a welcome from the Lord Mayor, and the presentation of prizes to the regional winners of national competitions in mathematics and physics—four young high school graduates with promising careers ahead of them. This was followed by a short, lighthearted dramatised account of Green's life and work. The Lord Mayor then hosted a Civic Luncheon in the Council House for academic guests and members of the Green family, while others adjourned to the University. In the afternoon came one of the most important events of the Celebrations: the delivery of Public Lectures on the invitation of the University by two of the most eminent scientists of this century, Professor Julian Schwinger, Nobel Laureate, of the University of California at Los Angeles, and Professor Freeman Dyson, F.R.S., of the Institute of Advanced Study, Princeton University, Princeton, New Jersey.

Scientists in the University of Nottingham knew that Julian Schwinger had been responsible for introducing Green's functions into quantum physics but were curious to know more. At the age of thirty, Schwinger had been hailed 'as one whom American physicists regard as the heir-apparent to the mantle of Einstein in producing the most important development in the last twenty years in our basic understanding of the cosmic forces holding the material universe together'.[15] So the University of Nottingham, prompted by Professor Challis, invited him to give a lecture on the subject. To the delight and satisfaction of all, Professor Schwinger duly arrived from Los Angeles, accompanied by his wife Clarice. This was a great honour since he was already in his mid-seventies: in fact Julian Schwinger died a year later almost to the day, on 16 July 1994, aged seventy-six. In these circumstances, one wondered whether this was the last lecture Julian Schwinger delivered. A posthumous appreciation (see Martin and Glashow, References II) of his work which appeared in 1995, 'Julian Schwinger, Prodigy, Problem Solver, Pioneering Physicist', contains the passage: 'As Schwinger noted in his last paper, *A Tribute to George Green* [Appendix VIa], the utilisation of Green's functions has been central to his work, to his students, and to physics in general'.[16] Mrs Clarice Schwinger believes that this was indeed so and adds: 'I don't remember when or if he gave any earlier talks in the U.K. I think not. He surely did not when he was involved in the Open University series.'[17] So as this does appear to have been Julian Schwinger's last public appearance, and in the light of his illustrious career, Green enthusiasts felt doubly honoured that it should have been in celebration of the bicentenary of George Green.

Schwinger laced an inevitably technical and specialised lecture with his dry wit, giving as his title: 'The Greening of Quantum Field Theory: George and I'. His account provided a detailed explanation of how he had come to appreciate that Green's functions had a natural place in quantum field theory. As such, it is a historic document and is reproduced in full, together with Freeman Dyson's equally informative lecture, in Appendix VI.

Freeman Dyson is known to physicists worldwide for his seminal work in quantum electrodynamics, and his name is linked with the three Nobel Laureates of 1965: Tomonaga, Schwinger and Feynman. Indeed Sylvan Schweber, in his book *Q.E.D. and the Men Who Made It: Dyson, Feynman, Schwinger and Tomonaga*, states: 'Yet I would suggest that Dyson should have shared the Nobel Prize, that his contributions recast the way field theories were thought of and dealt with, and that his criterion of renormalizability for selecting theories had great import on the subsequent evolution of the subject. Without minimizing in any way Tomonaga's accomplishments, it seems to me that the developments in the period from 1947 to 1950 would not have been substantially different without him, yet one cannot conceive of the subsequent developments without Dyson'. However, Schweber adds a note: 'Only three people can share in the award, which made the assignments particularly difficult'.[18]

Dyson paid his tribute to George Green in his lecture 'Homage to George Green: How Physics Looked in the Nineteen-Forties'. It formed a perfect complement to Schwinger's account, since it was not one man's personal odyssey but a wide-ranging view of the development of quantum field theory in America, from the time of Dyson's arrival at Cornell University, as a research student fresh from a Junior Fellowship at Cambridge in 1948, up to the nineteen-eighties. In the course of his account, Dyson charted the increasing use of Green's functions in the development of particle physics although, as he points out, they were not always labelled as such.

After the lectures, the University paid a further tribute to Green in a short ceremony in which the Science Library was re-named the George Green Library of Science and Engineering. Some time later a Green Mural was commissioned by the Library and accepted on behalf of the University by Professor Challis, who was then Pro Vice-Chancellor. It is an arresting work, occupying a long wall over a staircase spanning three floors. As ever, the lack of any likeness of George Green posed a problem, which was skilfully solved by the artist Derek Hampson. The final version of the work combined naturalistic and symbolic features, in which a horizontal sequence of yellow, green and sepia suggested the passing seasons of the year. These merging bands of colour are overlaid by a shower of brightly coloured autumn leaves representing the transitory nature of human life. The whole is superimposed on an unobtrusive background of squares, each with its own subtly distinctive pattern. In the words of the artist: 'Through these coloured grids, lines representing particles propagate, or move, from corner to corner of the squares. The concept of this was that they represent particles in motion, the particles that Green's work was so concerned to measure'. The impact of the mural, which aimed to represent in visual terms 'Green's highly complex and abstract work and his life, which was short and not fully documented', is both challenging and aesthetically pleasing. In fact some percipient viewers of the mural claim to have discerned the faint, almost subliminal, outline of a mill spanning its full length—although this might be a personal response to the evocative nature of Hampson's work. Science students and staff intrigued by the mural will find information about Green, and a colour photograph of the restored Green's Mill, on an adjacent wall.

The following morning, Professors Schwinger and Dyson attended a degree ceremony in which they were awarded honorary degrees of the University of Nottingham. They and members of the University later travelled to Caius College, Cambridge, where they were welcomed by Sir Sam Edwards. After tea in the Fellows' Parlour, they met other guests, including Martin Neary, Organist at Westminster Abbey, and Lady Jeffreys, widow of Sir Harold Jeffreys (who in 1924 anticipated the W.K.B. method).[19] They were then taken to view windows in College Hall, in which six scientists, all associated with Caius, had been commemorated. The central light of each window showed a diagrammatic representation in stained glass of some significant feature of the work in which each scientist had been engaged. In George Green's case, this was described as 'his eponymous theorem in the three-dimensional calculus . . .commemorated in a diagram, necessarily two-dimensional, capturing the essence of the theorem'. The account by Dr A.W.F. Edwards, Fellow and informal chairman of those associated with the project, added: 'Green's window caused the most discussion because of the difficulty of rendering Green's Theorem in two dimensions without it becoming Stokes' Theorem'.[20] Later the Master of Caius hosted a dinner held in the Fellows' Combination Room, which had been the Library in Green's time.

The last day of the Celebrations was in London, starting with a Bicentennial Commemoration at the Royal Society. This was yet another highly significant tribute to Green, following upon those offered by his native city and his Cambridge college. A Fellowship of the Royal Society is one of the highest honours a scientist can receive. But it was not always so, and the story of Green's life reveals why none of his papers was published in the Society's *Transactions*.[21] Given the high prestige enjoyed by the Society today however, it was with pride and deep appreciation that the Green enthusiasts collaborated in a programme organised by one of their number, Professor A.J.M. Spencer, F.R.S. This was in consultation with representatives of the Royal Society, the Institute of Physics, the London Mathematical Society, the Institute of Mathematics and Its Applications and the Institute of Electrical Engineers, all of whom were represented at the Dedication Service.

After a welcome from the President, Sir Michael Atiyah, and an Introduction by Professor Spencer, Professor Challis introduced the speakers of the morning session. The first paper was, perhaps inevitably, in view of the previous lack of knowledge concerning Green's life, an account by his biographer and Secretary of the George Green Memorial Fund, Mary Cannell. There followed a paper on 'Surface Integrals and Divergence Theorems: French Mathematics and Green, 1824–1828' by Dr. Ivor Grattan-Guinness of Middlesex University and an authority on this topic and period. The morning concluded with a shortened version of his Nottingham lecture by Professor Freeman Dyson, 'Homage to George Green: How Physics Looked in the Nineteen-Forties'. The afternoon session, chaired by Professor Spencer, was devoted to contemporary aspects of Green's contribution to science. Professor David Sherrington, of Oxford University, spoke on 'Green's Functions in Solid State Physics', and Professor D.S. Jones, F.R.S., of the University of Dundee, on 'Green's Functions and Electromagnetism'. Professor Sir James Lighthill, F.R.S., of University College, London, gave a paper on 'Aeroacoustical Uses of Green's Functions—Noise Emission from Jets and Other Flow Systems', and the last paper, 'A Trace from George Green to Bifurcation', was delivered by Professor Dr K. Kirchgässner of the Universität Stuttgart. These later papers showed the continuing vitality of Green's methods, which have become almost a lingua franca amongst engineers and scientists. They revealed that Green's fundamental insight—that using Green's functions to solve problems provides not just quantitative information about energies, but also conceptual information about the nature of the interactions involved—has reaped dividends in a wide and varied range of applications.[22]

The national accolade accorded to the greatest minds of Britain, of the past and recent present, is commemoration in Westminster Abbey. 'Poets' Corner' in the East Transept is a well-established attraction for visitors. Less well known is the area dedicated to men of science on the far side of the Sanctuary, of which the focal point is the memorial to Isaac Newton. His statue stands on the left side of the Quire, with his grave below. Nearby are the graves of William Herschel and his famous son (and friend of Edward Bromhead) John Frederick Herschel. A medallion with a profile portrait of George Gabriel Stokes is embedded nearby in the back of the Quire wall. Around Newton's grave are diamond-shaped plaques inscribed with the names of Michael Faraday, James Clerk Maxwell and Kelvin. Elsewhere in the Sanctuary are stone slabs bearing the names of Lord Rayleigh, J.J. Thomson and Ernest Rutherford. These are illustrious names from the nineteenth and early twentieth centuries. George Green, who was actually born in the eighteenth century, was the first scientist to be commemorated for over sixty years; in 1995 Paul Dirac was similarly honoured.[23]

The site of the plaque to Green was swiftly agreed upon. The Abbey authorities suggested it be placed near Newton's grave, in the company of those to Kelvin, Faraday and Clerk Maxwell. The site was not only prestigious, it was also appropriate. Whittaker had polarised the names of Newton and Green when he declared that the latter's scientific activity had brought to an end the darkest period in the history of the University (i.e., Cambridge) since the death of Newton.[24] It was Kelvin who re-discovered the Essay of 1828 and established Green's place in nineteenth century science, and it was Clerk Maxwell who developed the wide field of electromagnetism, taking the theoretical contributions of Green and combining them with the experimental discoveries of Faraday, who was also Green's great contemporary.

The inscription on the plaque took longer to decide upon. The original choice was 'George Green: Miller, Mathematician and Physicist 1793–1841'. This represented the wishes of various groups on the Green committee. An early suggestion had been 'Mathematical Physicist'—a term often used in relation to Green—but the physicists contended that Green's contributions to physics were no less than his contributions to mathematics and that separate terms should be used. A further consideration was that the restoration of Green's Mill had been loyally supported by the people of Sneinton, who felt that Green's association with a now quite famous city landmark should be recognised, so 'Miller' was also included in the attribution. However, in further discussions, it was pointed out that Green was finally being honoured for his achievements as a mathematician and physicist, and not as a miller. The inclusion of the latter term was in fact ironic, since Green had deeply disliked his milling as being a hindrance to his mathematical studies. Finally, an agreement was reached which was both felicitous and aesthetic. The former inscription in any case had sat uneasily within the confines of the diamond-shaped plaque. So 'Miller' was omitted but a bronze roundel, depicting Green's Mill in perfect detail, was inserted in the apex—a suitable reminder for those who know Green's background and a cause for enquiry from those who do not.

The Dedication Ceremony took place in the Nave of the Abbey and followed Evensong which is sung daily in the Quire. It opened with the Procession, in full academic dress, of representatives of the major British scientific societies who have been associated with Green's scientific work: the President of the Royal Society and Master of Trinity College Cambridge, Sir Michael Atiyah; Sir William McCrea, Senior Fellow at 91, of the Royal Society of Edinburgh, (the President being too ill to attend); Professor Peter Gray, Master of Gonville and Caius College, Cambridge and President of the Cambridge Philosophical Society; Professor Sir Sam Edwards, President of Gonville and Caius College and Cavendish Professor of Physics; the President of the Institute of Physics, Mr C.A.P. Foxell; Professor J.R. Ringrose, President of the London Mathematical Society; Professor M.J. Rycroft, Honorary Secretary of the Institute of Mathematics and Its Applications; Mr John Fauvel, President of the British Society for the History of Mathematics; Professor Colin Campbell, Vice-Chancellor of the University of Nottingham; Professor L.J. Challis, Lancashire-Spencer Professor of Physics at the University of Nottingham and Chairman of the George Green Memorial Fund; and its Honorary Secretary, Miss D.M. Cannell. George Green's international reputation was reflected in the presence of the two prestigious guests, Professor Julian Schwinger and Professor Freeman Dyson.

Those who knew the George Green story recalled his declining Bromhead's invitation to visit Cambridge to see his friends and 'others who constitute the Chivalry of British Science. Being as yet only a beginner I think I have no right to go there and must defer that pleasure until I shall have become tolerably respectable as a man of science should that day ever arrive'. They were now convinced that in the presence of a scientific gathering such as this, that day had indeed arrived.

Many eminent scientists were among the large congregation seated in the Nave, including a number of Past Presidents of the Institute of Physics. A special visitor for Freeman Dyson was Professor Nicholas Kemmer, his Cambridge tutor and mentor, who attended the Celebrations, and to whom he had paid grateful tribute in his lecture in Nottingham and that morning at the Royal Society.[25] Civic dignitaries were also present, led by the Lord Mayor of Westminster, and the Lord and Lady Mayoress of Nottingham, as well as the Sheriff and his Lady. The pews behind them were filled by a large number of Green enthusiasts, including the Chairman and Secretary of the Bromley House Library, members of the Sneinton Environmental Society and many others from Nottingham and elsewhere.

The Abbey clergy, led by the Dean and including the Precentor and several Canons, together with those participating in the ceremony, sat in the Sanctuary facing Newton's grave and the plaque to George Green, covered with a veiling cloth of dark green velvet bearing the Arms of the City of Nottingham, worked in lace, the town's principal industry in Green's time. The Green descendants sat in a separate group in the Sanctuary, opposite the clergy and those taking part in the service. On this occasion, it was Mrs Marjorie Carey, the late Mrs Saunders' younger sister who, as the last remaining great-granddaughter, provided the nearest link with the mathematician: she was accompanied by three of her children, Maurice and his sisters, Mrs Bridget Broom and Mrs Ann Atherton. Her nephew and niece, Mr Brian Voss and Mrs Jacqueline Bigg, arrived from Canada. Mrs Saunders' son, Mr George Saunders, and grandson, Dr Christopher Andrews, were also present. The two latter had attended the celebrations at the opening of the restored mill in 1985, as had the two members representing the earlier family of George Green Moth, Mr James Hall and his sister, Mrs Elizabeth Draper. She was joined for the Abbey ceremony by her two children, David and Janine, the latter having come over from New York for the ceremony.

Professor Challis, Chairman of the George Green Memorial Fund, opened the Service by reading extracts relating to Green's scientific work, taken from the writings of Edward Bromhead, Harvey Goodwin, Kelvin, Stokes and Robert Schrieffer, Nobel Laureate, this last from a letter written in 1974.[26] There followed extracts from Green's letters to his patron Bromhead, read by Mary Cannell, Secretary of the Fund. It was between these two readings that the Abbey music reflected a personal element in that Martin Neary, the Abbey Organist, had been a pupil of Freeman Dyson's father, Sir George Dyson.[27] At Evensong, the Versicles and Responses were sung to a setting by Martin Neary, who later conducted the Magnificat and Nunc Dimittis to Dyson's splendid setting in D. In Freeman Dyson's autobiography, *Disturbing the Universe*, there is a chapter titled 'Prelude in E-Flat Minor' in which he writes: 'At an early age I found my father's copy of Bach's forty-eight Preludes and Fugues for the well-tuned piano, and studied carefully the arrangements of sharps and flats in the key signatures. My father explained to me how Bach worked his way twice through all the twenty-four major and minor keys. But why is there no prelude in E-flat minor in the second book? My father did not know . . .'. A page or two later he adds 'The Prelude in E-flat minor continued to be my favourite'; Dyson then records how in later life, he heard the Prelude, 'his old friend of long ago. Superbly played, played just the way my father used to play it'—but in totally unexpected circumstances. This experience brought reconciliation following a difficult professional and personal episode, and restored his piece of mind.[28] In homage to father and son, Neary played the piece on a piano specially brought into the Nave. Following the readings, Sir Michael Atiyah, President of the Royal Society, unveiled the plaque and asked the Dean of Westminster 'to receive into the safe custody of the Dean and Chapter, here in the Nave of the Abbey among the memorials of other scientists, this memorial in honour of the mathematician and physicist, George Green' after which Professor Sir Sam Edwards gave the Address. The service concluded with prayers, including one by James Clerk Maxwell, and members of the Green family placing flowers round the plaque.

The Celebrations finally ended with a reception on the Terrace of the House of Commons in the Palace of Westminster (more often referred to as the Houses of Parliament). This was another privileged occasion. Such an event is by invitation only and is hosted by the Member of Parliament (M.P.) in whose constituency the celebrating organisation is based. He or she must have trust in the probity and monetary standing of the body concerned, since at the end of the day, it is the M.P. who is financially responsible! For guests, this is a unique experience, since the Commons Terrace—the other half is the preserve of the House of Lords—is on a balcony over the Thames.

The view of the river in the fading light, with the water reflecting the setting sun, is unforgettable.

Thus the Celebrations in honour of George Green's Bicentenary came to a close, and finally sealed his posthumous reputation. This was a far cry from the modest obituary in the Nottingham Review of 1841: 'Had his life been prolonged, he might have stood eminently high as a mathematician'. At this point, the writer is irresistibly reminded of a later passage in the paper written by John Toplis in 1805 which has been quoted earlier: 'It is possible that discoveries more wonderful and of greater utility than those already made by the help of mathematics, may some time or other be effected, should some great genius once point out the way'.[29] Would it be conceded that, apart from his other contributions to science, Green's functions represent one of those discoveries? If so, and if the hypothesis advanced earlier in this narrative is correct, it is ironic to think that Toplis himself may have tutored just such a genius.[30]

This viewpoint, however, is retrospective. The impetus for commemorating the bicentenary of George Green did not come from a nostalgic—or guilty—awareness of belated recognition. It came from the fact that Green's mathematics has made a 'quantum leap' into this century. Green's contribution to science was recognised in the nineteenth century, where it was firmly established in classical physics, but his position is now even more firmly established in the quantum age. Green's functions are the valued mathematical tool used by quantum physicists worldwide. It was largely this factor, as well as the use of his eponymous Theorem, and the wide and varied applications of his scientific concepts to modern physics and technology, that have singled out Green from his nineteenth century peers, and occasioned this long-deserved celebration of his life and work. His future would now appear secure. In the words of Julian Schwinger: 'What finally shall we say about George Green? Why he is, in a manner of speaking, alive, well and living among us'.

APPENDICES

Appendix I

The Mathematics of George Green

by M.C. Thornley

This appendix concentrates on Green's first publication, 'An Essay on the Application of Mathematical Analysis to the Theories of Electricity and Magnetism', as it contains work which was both original and of his own devising. The later publications also contain original work to a lesser degree, but were to some extent influenced by the suggestions of Bromhead, and by Cambridge luminaries.

Although Green chose to apply mathematics to physical situations and not to develop some mathematics for its own sake, it was in the development of mathematical ideas to solve specific physical problems that he achieved the successes for which he is primarily remembered. His stated object in the Essay was to 'submit to Mathematical Analysis the phenomena of the equilibrium of the Electric and Magnetic Fluids, and to lay down some general principles equally applicable to perfect and imperfect conductors' (N.F. 9). As discussed in the main text, the origins of his interest and expertise in mathematics and its applications can only be surmised.

Green divided the Essay into five parts, the last three being further divided into numbered sections:

Preface (N.F. 3–8)
Introductory Observations (N.F. 9–18)
General Preliminary Results, Articles 1–7 (N.F. 19–41)
Application of the Preceding Results to the Theory of Electricity, Articles 8–13 (N.F. 42–82)
Application of the Preceding Results to the Theory of Magnetism, Articles 14–17 (N.F. 83–115)

The following original concepts appear in the text – the first in the

'Introductory Observations', the rest in the 'General Preliminary Results':

(i) the name 'potential function' (N.F. 9);
(ii) the theorem now known as Green's Theorem (N.F. 23);
(iii) the idea of reciprocity (N.F. 26);
(iv) the term 'singular value' (N.F. 27);
(v) the technique now known as Green's functions (N.F. 32).

Formal proofs had to await later analysts; Green based his proofs on the needs of the physical situation he was investigating. The printed text shows a couple of misprints; no mathematical error has been found in the book.

(i) Potential function

> It is well known, that nearly all the attractive and repulsive forces existing in nature are such, that if we consider any material point p, the effect, in a given direction, of all the forces acting upon that point, arising from any system of bodies S under consideration, will be expressed by a partial differential of a certain function of the co-ordinates which serve to define the point's position in space. The consideration of this function is of great importance in many inquiries, and probably there are none in which its utility is more marked than in those about to engage our attention. In the sequel we shall often have occasion to speak of this function, and will therefore, for abridgement, call it the potential function arising from the system S. (N.F. 9).

Green went on to define the function in verbal terms, which he then put into mathematical terms in the next part. Following the French examples, he used the sign $\int$ for single, double and triple integrals, and the sign d both for derivatives and for partial derivatives. Thus he wrote:

> Firstly, let us consider a body of any form whatever, through which the electricity is distributed according to any given law, and fixed there, and let x', y', z', be the rectangular co-ordinates of a particle of this body, ρ' the density of the electricity in this particle, so that $dx'dy'dz'$ being the volume of the particle, $\rho'dx'dy'dz'$ shall be the quantity of electricity it

contains: more-over, let r′ be the distance between this particle and a point p exterior to the body, and V represent the sum of all the particles of electricity divided by their respective distances from this point, whose co-ordinates are supposed to be x, y, z, then shall we have

$$r' = \sqrt{(x'-x)^2 + (y'-y)^2 + (z'-z)^2} ,$$

and

$$V = \int \frac{\rho' dx' dy' dz'}{r'} ;$$

the integral comprehending every particle in the electrified mass under consideration.

LAPLACE has shown, in his Méc. Céleste, that the function V has the property of satisfying the equation

$$0 = \frac{d^2V}{dx^2} + \frac{d^2V}{dy^2} + \frac{d^2V}{dz^2} ,$$

and as this equation will be incessantly recurring in what follows, we shall write it in the abridged form $0 = \delta V$; the symbol δ being used in no other sense throughout the whole of this Essay. (N.F. 19, 20)

On the next page (N.F. 21), Green writes: 'Let now q be any line terminating in the point p, supposed without the body, then $-\frac{dV}{dq}$ = the force tending to impel a particle of positive electricity in the direction of q, and tending to increase it.'

Thus Green knew that his potential function V satisfied Laplace's Equation $\nabla^2 V = 0$ (in modern notation) and that minus its space rate of change gave the value of the force acting in that direction. He did not link the word 'energy' with the function V. Whether he meant that V was a function 'of great potential' in the mathematics which followed, or whether he considered it as a measure of potential energy, is not known – suffice it to say that he invented a name which has stood the test of time.

(ii) Green's Theorem

Before proceeding to make known some relations which exist between the density of the electric fluid at the surfaces of bodies, and the corresponding values of the potential functions within and without those surfaces, the electric fluid being confined to them alone, we shall in the first place, lay down a general theorem which will afterwards be very useful to us. This theorem may be thus enunciated;

Let U and V be two continuous functions of the rectangular co-ordinates x, y, z, whose differential coefficients do not become infinite at any point within a solid body of any form whatever; then will

$$\int dxdydz\, U\, \delta V + \int d\sigma\, U \frac{dV}{dw} = \int dxdydz\, V\, \delta U + \int d\sigma\, V \frac{dU}{dw};$$

the triple integrals extending over the whole interior of the body, and those relative to $d\sigma$, over its surface, of which $d\sigma$ represents an element: dw being an infinitely small line perpendicular to the surface, and measured from this surface towards the interior of the body. (N.F. 23)

Thus, for his stated initial conditions, Green established a connection between the summation effects through the interior of a body with the summation effects over the body surface. Green's Theorem is also known as Green's Second Identity and is equivalent to the Divergence Theorem.[1] A version for a plane exists, which links the summation effects through an area with the summation effects round the boundary.[2]

Hence the summation effects over a surface can be obtained by computing the summation effects through the volume, or vice versa. A difficult – or seemingly impossible – computation can thus be made by working out an easier, or possible, equivalent computation. Also, if three of the terms can be computed, then the fourth can be found, which is sometimes easier than trying to find the required result directly. Green developed the Theorem for this second reason.

In more modern notation, Green's Theorem can be written

$$\iiint U\, \nabla^2 V\, dv + \iint U \frac{\partial V}{\partial n} d\sigma = \iiint V\, \nabla^2 U\, dv + \iint V \frac{\partial U}{\partial n} d\sigma,$$

where Green's dxdydz, w are replaced by dv, n respectively.

Green used the symbol $\int$ for all types of integration until his last two papers of 1839, when he adopted double and triple integral signs without comment. No distinction of symbol was ever made between ordinary and partial differentials, and vector notation was not available in his time.

Green's use of Green's Theorem will be shown below.

Green began his Introductory Observations by defining the potential function and then proceeding with a general statement of the problem he proposed to tackle:

> Suppose it were required to determine the law of the distribution of the electricity on a closed conducting surface A without thickness, when placed under the influence of any electrical forces whatever: these forces, for greater simplicity, being reduced to three, X, Y, and Z, in the direction of the rectangular co-ordinates, and tending to increase them. Then ρ representing the density of the electricity on an element $d\sigma$ of the surface, and r the distance between $d\sigma$ and p, any other point of the surface, the equation for determining ρ which would be employed in the ordinary method, when the problem is reduced to its simplest form, is known to be
>
> $$\text{cons} = a = \int \frac{\rho d\sigma}{r} - \int (Xdx + Ydy + Zdz) \;\ldots\ldots\; (a)\;;$$
>
> the first integral relative to $d\sigma$ extending over the whole surface A, and the second representing the function whose complete differential is $Xdx + Ydy + Zdz$, x, y and z being the co-ordinates of p. (N.F. 10)

Having stated the 'standard approach' available to him, he then states his objections to it, which led him to develop the Theorem:

> This equation is supposed to subsist, whatever may be the position of p, provided it is situate upon A. But we have no general theory of equations of this description, and whenever we are enabled to resolve one of them, it is because some consideration peculiar to the problem renders, in that particular case, the solution comparatively simple, and must be looked upon as the effect of chance, rather than of any regular and scientific procedure. (N.F. 10)

Green had noted in the preface:

> The advantages LAPLACE had derived in the third book of the *Mécanique céleste*, from the use of a partial differential equation of the second order, there given, were too marked to escape the

> notice of any one engaged with the present subject, and naturally served to suggest that this equation might be made subservient to the object I had in view. (N.F. 7)

He thus acknowledges the importance of the Laplace Equation $\nabla^2 V = 0$ to his work; the term 'second order partial differential equation' establishes the type of equation in words, but the symbolism to distinguish it from an ordinary second order differential equation was lacking in Laplace, and Green presumably saw no reason to eliminate possible ambiguities. That Green was primarily a mathematician, whilst working in what was later to be known as physics, is shown by his adept sidestepping of the problems of fluid theory, and by the complete lack of diagrams in his publications.

He showed that his potential function V satisfied Laplace's Equation $0 = \nabla^2 V$ for points exterior to the body, but that for points within the body the equation had to be modified to $0 = \nabla^2 V + 4\pi\rho$, or, in his notation, '$0 = \delta V + 4\pi\rho$ ..... (1)' (N.F. 22), where ρ is the density of electricity within the body. He did this by considering the contributions of an exceedingly small sphere enclosing the point in question together with that of the rest of the body exterior to the sphere. Equation (1) is known today as 'Poisson's Equation', as his was the work which became generally acknowledged.

As the Essay was the final polished product, Green gives no hints as to what led him along the path to the Theorem, and as there seem to be no surviving notebooks, the likelihood of our ever knowing must be remote.

Green proved his Theorem by considering the triple integral

$$\int dxdydz \left\{\left(\frac{dV}{dx}\right)\left(\frac{dU}{dx}\right) + \left(\frac{dV}{dy}\right)\left(\frac{dU}{dy}\right) + \left(\frac{dV}{dz}\right)\left(\frac{dU}{dz}\right)\right\}.$$

By using integration by parts and an appeal to the values of the quantities on the surface of the body, he was able to show that the integral was equivalent to

$$-\int d\sigma\, V\frac{dU}{dw} - \int dxdydz\, V\,\delta U \qquad \text{(N.F. 24, 25).}$$

In modern notation this equivalence can be written

$$\int_V (V\,\nabla^2 U + \underline{\nabla} U.\underline{\nabla} V)\, dv = \oint_S V\frac{\partial U}{\partial n}\, da,$$

when it is known as Green's First Identity.

He could have proceeded through the sequence of manipulations once again, substituting V for U and U for V to arrive at a second equivalent answer,

$$- \int d\sigma \, U \frac{dV}{dw} - \int dxdydz \, U \, \delta V,$$

and perhaps he did, but instead he used RECIPROCAL RELATIONS for the first known time in mathematical physics: 'since the value of the integral . . . remains unchanged when we substitute V in the place of U and reciprocally, it is clear, that it will also be expressed by

$$- \int d\sigma \, U \frac{dV}{dw} - \int dxdydz \, U \, \delta V \text{'} \qquad \text{(N.F. 25, 26)}.$$

Equating these two pairs of integrals gives Green's Theorem, in which he altered the order of terms compared with the initial statement:

$$\int d\sigma \, V \frac{dU}{dw} + \int dxdydz \, V \, \delta U = \int d\sigma \, U \frac{dV}{dw} + \int dxdydz \, U \, \delta V \; \ldots \; (2)$$

> . . . In our enunciation of the theorem, we have supposed the differentials of U and V to be finite within the body under consideration . . . which is understood in the method of integration by parts there employed. In order to show more clearly the necessity of this condition, we will now determine the modification which the formula must undergo, when one of the functions, U for example, becomes infinite within the body . . . in one point p′ only: (N.F. 26).

He showed that the right-hand side of equation (2) had an extra term, $-4\pi V'$, where V' is the value of V at p′.

That he had problems in accomplishing this result can be inferred from the preface: 'Recollecting, after some attempts to accomplish it, that previous researches on partial differential equations, had shown me the necessity of attending to what have, in this Essay, been denominated the SINGULAR VALUES of functions . . .' (N.F. 7).

By placing a small sphere of radius a round p′, he showed that equation (3) would hold for the rest of the body (with greatest error of order a^2), and that the sphere would introduce a limiting contribution of $-4\pi V'$, giving the required result as

$$\int dxdydz\, U\, \delta V + \int d\sigma\, U \frac{dV}{dw} = \int dxdydz\, V\, \delta U + \int d\sigma\, V \frac{dU}{dw} - 4\pi V'$$

. . . (3) (N.F. 27).

If V becomes infinite at some point p″ within the body and U has value U″ at that point then the left-hand side of (2) has the additional term $- 4\pi U''$: 'The same process will evidently apply, however great may be the number of similar points belonging to the functions U and V ' (N.F. 27).

Green restates his name for points where a function becomes infinite: 'For abridgement, we shall in what follows, call these singular values of a given function, where its differential coefficients become infinite, and the condition originally imposed upon U and V will be expressed by saying, that neither of them has any singular values within the solid body under consideration' (N.F. 27).

Thus, during the proof of his Theorem, Green devised two terms which have endured.

(iii) Reciprocal relations, or reciprocity

This arises when there is an essential symmetry in a system or equation. A simpler illustration can be found in the use of a simple convex lens of focal length f. If an object distant u from the lens (u > f) gives rise to a sharp image distant v from the lens, then an object distant v from the lens will give rise to a sharp image distant u from the lens. Apart from experiment, this can be seen from the equation which connects u, v and f, namely $1/u + 1/v = 1/f$, in which u can be replaced by v, and reciprocally, and the equation is unchanged.

Green's opening paragraph to Article 6 is:

> Let now A be any closed surface, conducting electricity perfectly, and p a point within it, in which a given quantity of electricity Q is concentrated, and suppose this to induce an electrical state in A; then will V, the value of the potential function arising from the surface only, at any other point p′, also within it, be such a function of the co-ordinates of p and p′, that we may change the co-ordinates of p, into those of p′, and reciprocally, without altering its value. Or, in other words, the value of the potential function at p′, due to the surface alone, when the inducing

electricity Q is concentrated in p, is equal to that which would have place at p, if the same electricity Q were concentrated in p′. (N.F. 36)

This has given rise to a Reciprocity Theorem named after Green, although not explicitly stated by him:

If $e_1, e_2, \ldots, e_n$ are the charges on and $ø_1, ø_2, \ldots, ø_n$ are the potentials of a number of conductors in one equilibrium state, and $e'_1, e'_2, \ldots, e'_n$ are the charges on and $ø'_1, ø'_2, \ldots, ø'_n$ are the potentials of the conductors in a second equilibrium state, then

$$\sum eø' = \sum e'ø. \quad [3]$$

An analogue to this electrostatics theorem is also found in current electricity.

The following is a proof of Green's Reciprocity Theorem, as Green might have given it, making use of Green's Theorem to compute a surface effect when the volume effect is known.

Consider a number of conductors enclosed by a surface S, where R is a region of space which is bounded by S and the surfaces of the conductors and where the normals n to S and to the conductors are in R.

Rearranging Green's Theorem to obtain surface statements on the left side and space statements on the right side of the equals sign,

$$\iint \left(ø\frac{\partial ø'}{\partial n} - ø'\frac{\partial ø}{\partial n} \right) ds = \iiint \left(ø\nabla^2 ø' - ø'\nabla^2 ø \right) dv \ \ldots \ \text{(i)}$$

As ø and ø′ are potential functions they will satisfy Laplace's Equation, giving $\nabla^2 ø = 0$ and $\nabla^2 ø' = 0$.

Hence the right-hand side of equation (i) is zero, giving

$$\iint \left(ø\frac{\partial ø'}{\partial n} - ø'\frac{\partial ø}{\partial n} \right) ds = 0. \ \text{(ii)}$$

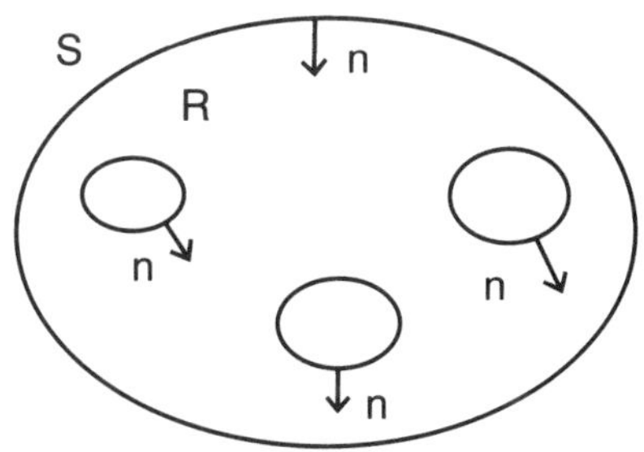

Rearranging equation (ii) and taking into consideration the separate surfaces,

$$\sum\iint \emptyset \frac{\partial \emptyset'}{\partial n} ds + \iint_S \emptyset \frac{\partial \emptyset'}{\partial n} ds = \sum\iint \emptyset' \frac{\partial \emptyset}{\partial n} ds + \iint_S \emptyset' \frac{\partial \emptyset}{\partial n} ds \ldots \quad \text{(iii)}$$

Apart from the direction of the normals, no restriction has been placed on the shape of S, which, for simplicity, will now be considered a sphere of radius r.

If r is very large, then both ø and ø′ are of the order r^{-1} and both $\partial \emptyset/\partial n$ and $\partial \emptyset'/\partial n$ are of the order r^{-2}, making both $\emptyset\partial \emptyset'/\partial n$ and $\emptyset'\partial \emptyset/\partial n$ of the order r^{-3}. ds is of the order $r^2 d\omega$, where $d\omega$ is an element of solid angle subtended by ds at the centre of the sphere.

Hence both $\iint_S \emptyset \frac{\partial \emptyset'}{\partial n} ds$ and $\iint_S \emptyset' \frac{\partial \emptyset}{\partial n} ds$ are of the order r^{-1},

and both tend to value zero as r becomes increasingly large.

In this case (iii) becomes $\sum\iint \emptyset \frac{\partial \emptyset'}{\partial n} ds = \sum\iint \emptyset' \frac{\partial \emptyset}{\partial n} ds$ .. (iv).

Now let ø and ø′ be the potential functions due to two distinct distributions of charge over the conductors, giving rise to charge densities of σ and σ′ respectively. As the electric force at the surface of a conductor is normal to it and equal to 4π times the charge density, then $-\partial \emptyset/\partial n = 4\pi\sigma$ and $-\partial \emptyset'/\partial n = 4\pi\sigma'$.

Thus equation (iv) becomes $\sum\iint \emptyset\sigma' ds = \sum\iint \emptyset'\sigma ds$ . . . (v).

As conductors are being considered, then ø and ø′ are constant over them, and equation (v) becomes $\sum \emptyset \iint \sigma' ds = \sum \emptyset' \iint \sigma ds$,

or $\sum \emptyset e' = \sum \emptyset' e$,

which is the result previously stated.

Green did use a similar technique (N.F. 29) in his work:

> If now, we conceive a surface inclosing the body at an infinite distance from it, we shall have, by applying the formula (2) of the same article to the space between the surface of the body and this imaginary exterior surface . . . the part due to the infinite surface may be neglected, because V′ is there equal to zero.

(iv) Singular values

In the previous paragraph, Green also stated: '. . . (seeing that here $\frac{1}{r} = U$ has no singular value) . . .' We have already noted that Green, 'after some attempts to accomplish it', had been unable to proceed with his work until he saw 'the necessity of attending to what have, in this Essay, been denominated the singular values of functions' (N.F. 7).

A function such as $U = \frac{1}{r}$ is defined for all values of r with the exception of r = 0. $U = \frac{1}{r}$ is not defined when r = 0, and some means must be made of excluding it, directly or indirectly, when working with the function. Green coined the term 'singular value' to describe this situation.

In some circumstances r = 0 will be excluded by the conditions of the problem. Green's imaginary exterior surface used only large values of r; hence r = 0 was automatically excluded – giving rise to his statement that '$\frac{1}{r} = U$ has no singular value'.

In other circumstances r = 0 is included in the conditions, and must then be given special consideration. Often a small 'exclusion zone' is placed around the singular point, and its contribution to the whole is considered separately from that of the rest of the region where the function is regularly behaved, having no singular points therein. Green used this technique when modifying his Theorem (2) to include a singular value of U within the body, thereby obtaining equation (3) (N.F. 26):

> We will now determine the modification which the formula (2) must undergo, when one of the functions, U for example, becomes infinite within the body; and let us suppose it to do so in one point p′ only: moreover, infinitely near this point let U be sensibly equal to $\frac{1}{r}$; r being the distance between the point p′ and the element dxdydz. Then if we suppose an infinitely small sphere whose radius is a to be described round p′, it is clear that our theorem (2) is applicable to the whole of the body exterior to this sphere, and since, $\delta U = \delta\frac{1}{r} = 0$ [i.e. $\nabla^2 U = \nabla^2\frac{1}{r} = 0$] within this sphere, it is evident, the triple integrals may still be supposed to extend over the whole body, as the greatest error that this supposition can induce, is a quantity of the order of a^2. Moreover, the part of $\int d\sigma\, U \frac{dV}{dW}$, due to the surface of the small sphere is only an infinitely small quantity of the order a; there only remains therefore to consider the part of $\int d\sigma\, V \frac{dU}{dW}$

due to this same surface, which since we have here

$$\frac{dU}{dw} = \frac{dU}{dr} = \frac{d\frac{1}{r}}{dr} = \frac{-1}{r^2} \; [4] = \frac{-1}{a^2},$$

becomes $-4\pi V'$ when the radius a is supposed to vanish.

Thus the only term in the Theorem (2) which is affected is the fourth one, which became two terms, $\int d\sigma\, V \frac{dU}{dW} - 4\pi V'$, the first due to all of the body except the small 'exclusion zone'; the second due to the effect of the 'exclusion zone' itself.

Singular values have, therefore, to be identified before a function is used and steps are devised to deal with them separately from the rest of the computation.

In the next Article, 4, Green found relations existing between the density of the electric fluid at the surface of the body and the potential functions then arising, both within (V at point p) and without (V′ at point p′) this surface. He showed that whilst both V and V′ satisfied Laplace's Equation, they were different functions which, none the less, had the same values at the surface, $\overline{V} = \overline{V'}$ (the horizontal lines indicating values at the surface). He used equation (3), together with an imaginary exterior surface, to show that

$$\rho = -\frac{1}{4\pi}\left\{\frac{\overline{dV}}{dw} + \frac{\overline{dV'}}{dw'}\right\} \quad .... \quad (4)$$

at a point on the surface, ρ being the surface density of the electricity (N.F. 30).

(v) Green functions, or Green's functions

It is in Article 5, tucked away in the second paragraph (N.F. 32), that Green's next important creation is to be found – a technique of solution for certain second order differential equations, later to be known as Green functions or Green's functions. We shall follow through his work and then comment on this powerful latent technique.

The aim of the Article was to find a formula which would give the value for the potential at any point p′, within the surface, when $\overline{V}$, its value at the surface itself is known; this Green found to be

$V' = -\iint \rho\overline{V}ds$, or, in his terms,

$V' = -\int d\sigma(\rho)\overline{V}$ ... (6), where ρ is the density that a unit of electricity concentrated in p′ would induce on this surface, if it conducted electricity perfectly and were put in communication with the earth.

Green started with the modified version of the Theorem, equation (3), and proceeded to eliminate the triple integrals from the statement:

> From what has been established (art. 3), it is easy to prove, that when the value of the potential function $\overline{V}$ is given on any closed surface, there is but one function which can satisfy at the same time the equation $0 = \delta V$ [i.e. Laplace's Equation] and the condition, that V shall have no singular values within this surface . . . by supposing $\delta U = 0$ . . .

Thus by showing that $\overline{V}$ and by assuming that $\overline{U}$ both satisfy Laplace's Equation, he was able to equate each triple integral in equation (3) to zero, leaving only three terms:

$$\int d\sigma\, \overline{U}\, \overline{\frac{dV}{dw}} = \int d\sigma\, \overline{V}\, \overline{\frac{dU}{dw}} - 4\pi V',$$

where the integrals are double integrals and the differential coefficients are partial differential coefficients.

'In this equation, U is supposed to have only one singular value within the surface, viz. at the point p′, and, infinitely near to this point, to be sensibly equal to $\frac{1}{r}$; r being the distance from p′ ' (N.F. 31).

In his Introductory Observations Green had written: 'But, as the general integral of the partial differential equation ought to contain two arbitrary functions, some other condition is requisite for the complete determination of V ' (N.F. 11). He needed to eliminate U from the equation which he had just obtained. He had already assumed that it satisfied Laplace's Equation; now he went on to impose further conditions:

'If now we had a value of U, which, besides satisfying the above written conditions, was equal to zero at the surface itself, we should have $\overline{U} = 0$ ' (N.F. 11). This meant that the term on the left-hand side of the equation became zero, leaving

$$0 = \int d\sigma\, \overline{V}\, \overline{\frac{dU}{dw}} - 4\pi V' \quad \dots \quad (5)$$

'which shows, that V′ the value of V at the point p′ is given, when

$\overline{V}$ its value at the surface is known' (N.F. 32).

The last stage needed was to appeal to the physical situation to show that a function such as U, with the conditions imposed, was possible, and to eliminate its partial derivative from equation (5):

> To convince ourselves that there does exist such a function as we have supposed U to be; conceive the surface to be a perfect conductor put in communication with the earth, and a unit of positive electricity to be concentrated in the point p′, then the total potential function arising from p′ and from the electricity it will induce upon the surface, will be the required value of U. For, in consequence of the communication established between the conducting surface and earth, the total potential function at this surface must be constant, and equal to that of the earth itself, i.e. to zero (seeing that in this state they form but one conducting body). Taking, therefore, this total potential function for U, we have evidently $0 = \overline{U}$, $0 = \delta U$, and $U = \frac{1}{r}$ for those parts infinitely near to p′. As moreover, this function has no other singular points within the surface, it evidently possesses all the properties assigned to U in the preceding proof.
>
> Again, since we have evidently $U' = 0$, for all the space exterior to the surface, the equation (4) art. 4 gives
>
> $$0 = 4\pi(\rho) + \frac{\overline{dU}}{dw} \quad ;$$
>
> where (ρ) is the density of the electricity induced on the surface, by the action of a unit of electricity concentrated in the point p′. Thus, the equation (5) of this article becomes
>
> $$V' = -\int d\sigma(\rho)\overline{V} \quad \ldots \quad (6)$$
>
> This equation is remarkable on account of its simplicity and singularity, [5] seeing that it gives the value of the potential for any point p′, within the surface, when $\overline{V}$, its value at the surface itself is known, together with (ρ), the density that a unit of electricity concentrated in p′ would induce on this surface, if it conducted electricity perfectly, and were put in communication with the earth. (N.F. 32)

In the above account Green placed three conditions on U, which is singular only at p′:

(i) $\overline{U} = 0$;

(ii) $\delta U = 0$, i.e. $\nabla^2 U = 0$;

(iii) $U = \frac{1}{r}$ for those parts infinitely near p′.

This 'restricted' definition of U is the first example of what Neumann first called Green's function.[6] Developed by later mathematicians, it has become fully generalized and is now a powerful tool for solving some types of partial differential equations.

The concept, as it is understood today, can be illustrated by the following example.

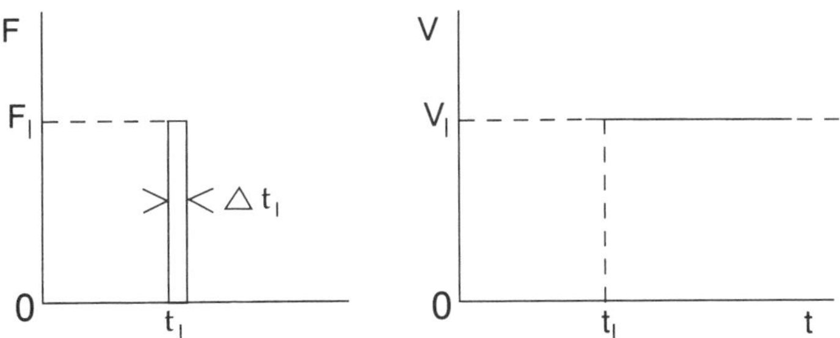

Consider a mass m at rest which is given an impulse P_1 at time t_1, due to a force F_1 being applied over a short time interval Δt_1. Then $P_1 = F_1\Delta t_1$. The ball was initially at rest, i.e. had zero momentum. After the blow it has acquired velocity v_1 and momentum mv_1. Thus $P_1 = mv_1 - 0$, or $F_1\Delta t_1 = mv_1$ and $v_1 = F_1\Delta t_1/m = P_1/m$.

Neglecting friction, etc., the mass will travel at constant velocity, v, due to the impulse, and will travel a distance x in a straight line. As the ball starts at time t_1, then at time t ($> t_1$),

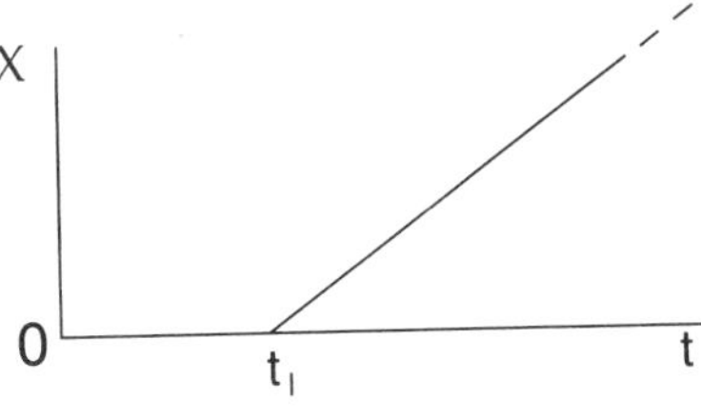

$$x = (t - t_1)v_1 = (t - t_1)\frac{P_1}{m} = \frac{(t - t_1)}{m} F_1\Delta t_1$$

We will DEFINE a Green's function, $G(t, t_1)$ as follows:

$$G(t, t_1) = \begin{cases} 0 & , \quad (t < t_1) \\ \dfrac{(t - t_1)}{m} & , \quad (t > t_1) \end{cases}$$

giving $x = G(t, t_1)\, F_1\Delta t_1$, or $x = G(t, t_1)\, P_1$.

Consider UNIT IMPULSE applied to the mass, i.e. $P_1 = 1$. Then

$$x = G(t, t_1).$$

$G(t, t_1)$ is thus a measure of the distance moved due to application of unit impulse; hence it has a physical connotation in the present case – namely, that of the response of the mass to the application of unit impulse at time t_1. For this reason a Green's function is sometimes also known as a RESPONSE FUNCTION. Its relationship to Green's own work will become clear in what follows.

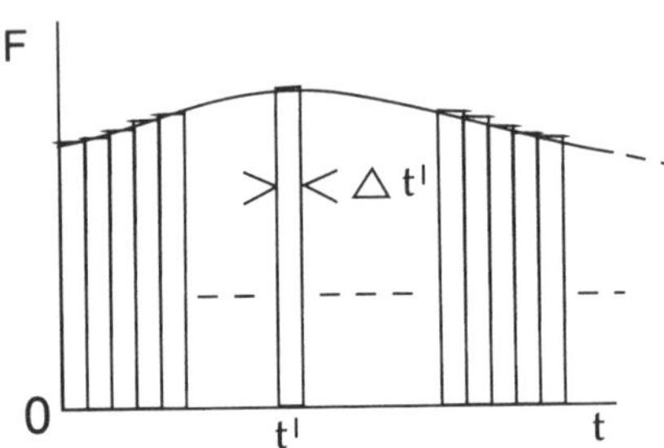

Consider a varying force acting on mass m, initially at rest. One approach is to consider the force providing a continuing succession of impulses. The distance moved will be given by

$$x = \left(\frac{t - t_1}{m}\right)F_1\Delta t_1 + \left(\frac{t - t_2}{m}\right)F_2\Delta t_2 + .. + \left(\frac{t - t_i}{m}\right)F_i\Delta t_i + .. \quad (t_1, t_2, ... < t)$$

$$= \sum_{i=1}^{n}\left(\frac{t - t_i}{m}\right)F_i\Delta t_i = \sum_{i=1}^{n} G(t, t_i)\, F_i\Delta t_i = \sum_{\Delta t'} G(t, t')\, F(t')\Delta t',$$

where $G(t, t') = 0$ for $t < t'$.

As $\Delta t'$ is made smaller and smaller, in the limit we get

$$x = \int_0^t G(t, t')\, F(t')dt'.$$

where $G(t, t')$ is a response function. If $F(t')$ is known in terms of t', then a solution may be possible.

e.g. for the case where $F(t') = F_0$ = constant, then

$$x = \frac{F_0}{m}\int_0^t (t - t')dt' = \frac{F_0}{m}\left[tt' - \tfrac{1}{2}t'^2\right]_0^t = \frac{F_0 t^2}{2m} = \tfrac{1}{2}at^2,$$

where a is the acceleration imparted by the force F_0.

Note that $G(t, t')$ is INDEPENDENT of the nature and composition of the force F. Therein lies much of its value.

A second example might help to highlight the potential usefulness of a Green's function. Consider a mass m oscillating vertically on the end of a helical spring, such that the spring does not become slack, with the initial conditions that the initial displacement is zero and the initial velocity is zero. The mass is subject to some forcing function F(t). The appropriate second order differential equation is $d^2x/dt^2 + k^2x = F(t)$, assuming that friction is negligible and that the spring constant is mk^2, the force required to produce unit extension. The standard method of solution is to solve the auxiliary equation, $d^2x/dt^2 + k^2x = 0$, to obtain the solution $x = A\sin(kt) + B\cos(kt)$, where A and B are arbitrary numbers. This is followed by finding a Particular Integral, whose form depends upon the function F(t).

An alternative method of solution is to find a Green's function for the problem. It will be a response function which incorporates the boundary conditions given. It is obtained by appealing to the Dirac delta function, when it can be found to be

$$G(t,t') = \begin{cases} 0 & , 0 < t < t' \\ \tfrac{1}{k}\sin[k(t-t')] & , 0 < t' < t \end{cases}$$

G(t, t') is independent of the nature of the forcing function, so that whatever F(t') may be, the solution will be

$$x = \int_0^t G(t, t')\, F(t')dt'.$$

The solution will, of course, depend on the nature of F(t'). If $F = F_0$ = constant, then

$$x = F_0\int_0^t \tfrac{1}{k}\sin[k(t-t')]\, dt' = F_0\left[\frac{-\cos[k(t - t')]}{k^2(-1)}\right]_0^t$$

$$= F_0(1 - \cos(kt))/k^2 ,$$

which is the result obtained by more conventional methods.

Green's functions for many second order differential equations plus boundary conditions have been tabulated for 'instant use'; otherwise they have to be constructed to fit the situation concerned.

Green's own 'Green's function' can now be stated in the following terms: If $\underline{r}'$ is the position vector of point p' and $\underline{r}$ is the position vector of any other point in the body or on its surface, then $G(\underline{r}, \underline{r}')$ is the total potential due to the perfectly conducting surface being earthed and a unit of positive electricity concentrated at p':

Green	Modern
$0 = \overline{U}$	$G(\underline{r}, \underline{r}') = 0$ if $\underline{r}$ is on the surface
$0 = \delta U$	$\nabla^2 G(\underline{r}, \underline{r}') = 0$ if $\underline{r} \neq \underline{r}'$
$U = \frac{1}{r}$	$G(\underline{r}, \underline{r}') \underset{\underline{r} \to \underline{r}'}{\sim} \frac{1}{\lvert\underline{r} - \underline{r}'\rvert}$

giving Green's equation (5) as

$$V(\underline{r}') = -\frac{1}{4\pi} \int_S \frac{\partial G(\underline{r}, \underline{r}')}{\partial n} V(\underline{r})\, ds.$$

A delta function is an infinitely sharply peaked function (e.g. an impulse due to a force over an infinitesimally small time interval), denoted by $\delta(x)$ such that

$$\delta(x) = \begin{cases} 0 & , x \neq 0 \\ \infty & , x = 0 \end{cases}$$

(Note that δ is used in a different sense to that used by Green.) To be meaningful the integral of $\delta(x)$ is normalized to unity, i.e.

$$\int_{-\infty}^{\infty} \delta(x) dx = 1.$$

Many important physical problems can be solved by – or simplified to a form which can be solved by – a second order differential equation: e.g. $\nabla^2 \o + f_1 \o = f_2$, where either or both f_1 and f_2 are constants or are functions of the co-ordinates whose form is known, over a domain D. Then the response or Green's function will be the solution of $\nabla^2 G + f_1 G = \delta(\underline{r}, \underline{r}')$, in the same domain D, subject to given boundary conditions, and where

$$\delta(\underline{r}, \underline{r}') = \begin{cases} 0 & , \underline{r} \neq \underline{r}' \\ \infty & , \underline{r} = \underline{r}' \end{cases} \text{, and } \int_D \delta(\underline{r}, \underline{r}')\, dr = 1.$$

By use of reciprocity and Green's Theorem we arrive at

$$\o(\underline{r}) = \oint_{S'} \left[\o(\underline{r}') \frac{\partial G(\underline{r}', \underline{r})}{\partial n'} - G(\underline{r}', \underline{r}) \frac{\partial \o(\underline{r}')}{\partial n'} \right] ds' + \int_{D'} f_2(\underline{r}')\, G(\underline{r}', \underline{r}) d\tau' \quad \ldots \text{(i)}$$

subject to whatever boundary conditions have been given. The method comes into its own when the boundary conditions are inhomogeneous – i.e. where the solution to be sought, or its normal derivative, is specified as a function of position on the boundary S of the domain D in which the solution is sought. For example, if ø is a specified function of position on the surface S, then $\partial \o / \partial n$ is not known. However, G can be chosen to be zero on S, giving also G′ zero on S′, making the second term zero and leaving only known quantities on the right-hand side of (i), and hence $\phi(\underline{r})$, provided G can be found and the integrals evaluated.

The method of Green's functions remained a mathematical curiosity in most cases until earlier this century, when its power was recognized and used. As well as in established topics such as mechanics, heat diffusion and electromagnetism, it has become a powerful tool in many other fields, including some aspects of quantum physics, chemical engineering, biochemical engineering, materials science, soil mechanics and nuclear engineering.

Green used his mathematical results in the rest of the Essay to solve various problems in static electricity and in magnetism. The first was to investigate Leyden phials charged in cascade (electric capacitors in series), and he came to the conclusion:

> that the total charge of all the phials is precisely the same, as that which one only would receive, if placed in communication with the same conductor, provided its exterior coating were connected with the earth. Hence this mode of charging, although it may save time, will never produce a greater accumulation of fluid than would take place if only one were employed. (N.F. 48)

This was possibly in response to arguments in Bromley House Library, or after one of Goodacre's lectures. Green had assumed no special shape for the phial, and went on to work out electrostatic effects for conductors of any shape, previous work having been restricted to particular shapes. He ended the electrical section by considering non-perfect conductors. The last four articles dealt with magnetism, founded on theories proposed by Coulomb and leading to results which were in agreement with those of Poisson, who had used a different method.

Green's work after the 'Essay'

After the 'Essay' Green wrote a further paper on electrostatics before moving on to deal with other topics. In this brief account, attention will be given to those papers in which he anticipated some techniques later to be developed for use in particular circumstances.

'On the Determination of the Exterior and Interior Attractions of Ellipsoids of Variable Densities' (Cambridge Philosophical Society *Transactions*, 1835)
Green used a law of attraction which was inversely proportional to the nth power of distance, associated with an unknown number of variables, s. He introduced a new quantity, u^2, to avoid 'violation of the law of continuity . . . in the values of the attractions when p passes from the interior of the ellipsoid into exterior space' (N.F. 187), and placed conditions on the values of the variables. He arrived at an integral

$$\int dx_1 dx_2 \ldots dx_s du\, u^{n-s} \left\{ \left(\frac{dV}{dx_1}\right)^2 + \left(\frac{dV}{dx_2}\right)^2 + \ldots + \left(\frac{dV}{dx_s}\right)^2 + \left(\frac{dV}{du}\right)^2 \right\};$$

'if . . . we endeavour to make' the above equation 'a minimum, we shall get in the usual way, by applying the Calculus of Variations, . . .' (N.F. 193).

This is the first known use of what Riemann called the Dirichlet Principle (that the function V which minimizes the above integral satisfies the Potential Equation).[7]

'On the Motion of Waves in a Variable Canal of Small Width and Depth. (Cambridge Philosophical Society *Transactions*, 1838)

> The equations and conditions necessary for determining the motions of fluids in every case in which it is possible to subject them to Analysis, have been long known, and will be found in the First Edition of the *Mécanique Analytique* of Lagrange. Yet the difficulty of integrating them is such, that many of the most important questions relative to this subject seem quite beyond the present powers of Analysis. There is, however, one particular case which admits of a very simple solution. The case in question is that of an indefinitely extended canal of small breadth and depth, both of which may vary very slowly, but in other respects quite arbitrarily. (N.F. 225)

Green arrived at a differential equation

$$0 = \frac{d^2\phi}{dx^2} + \left\{\frac{d\beta}{\beta dx} + \frac{d\gamma}{\gamma dx}\right\}\frac{d\phi}{dx} - \frac{1}{g\gamma}\frac{d^2\phi}{dt^2}$$

where ø is a function of x and t, 2β is the width, 2γ is the depth of the canal, and g is the acceleration due to gravity (the acceleration of free fall).

'It now only remains to integrate this equation' (N.F. 228). Green achieved this by making the assumption that β and γ were functions of x which were varying very slowly. He wrote $\beta = \gamma(\omega x)$ and $ø = A\,f(t + X)$, where ω is a very small quantity and A is a function of x of the same kind as β and γ. He substituted these into the above differential equation, neglecting terms of the order ω^2. Equating the co-efficients of f' and f'' to zero gave two solvable differential equations, which in turn gave rise to a solution for ø and also for u, the velocity of the fluid particles in the direction of x. He was also able to make statements about the wave's height, length and velocity.

The method he invented to solve the differential equation by using the first few terms of an asymptotically convergent series to obtain an approximate solution anticipated the Wentzel, Kramers and Brillouin (or W.K.B.) method by some ninety years.[8]

Study of one-dimensional wave mechanics produced a differential equation of the type $d^2\psi/dx^2 + f(x)\psi = 0$. The analytical theory of Green's approach was developed at the turn of the century and formalized by Jeffreys in 1924.[9]. It was also developed in turn and independently by Wentzel (1926), Kramers (1926) and Brillouin (1926)[10] in an attempt to relate the quantum theories of Bohr and de Broglie with the wave-mechanical theory of Schrödinger.[11] f(x) is assumed to be a slowly varying function of x, and ψ, the function to be found, is written $\psi = A(x)\,f(t + X)$, where X is a function of x and A(x) is also assumed to be slowly varying. Substitution into the differential equation and neglecting very small quantities gives a solvable equation whose solution is a good approximation to the solution of the original problem.

'On the Propagation of Light in Crystallized Media' (Cambridge Philosophical Society *Transactions*, 1839)
Although Green's topic was the propagation of light, much of this work had to do with elasticity, where deformations vanish when the deforming force is released. Tensors were unknown to Green; it is

a tribute to his pioneering work that one tensor used in elasticity theory is called Green's Tensor (tensor means 'stretched' – it is also the name given to a muscle which stretches a part).

For example, seismic sources may give rise to an equation which can be solved by use of a Green's function, which, in fact, is also a Green's tensor. An electromagnetic field can be expressed exclusively either in terms of components of electric intensity or in terms of components of magnetic intensity by the introduction of an appropriate Green's tensor.[12]

Green showed great mathematical ability and a sound insight into the ways in which his mathematics could be applied to physical situations; he was able to bring original thinking to bear on what he saw.

His later papers were possibly the subject of some argument with his contemporaries, as he included the statement:

> Perhaps I may be permitted on the present occasion to state, that though I feel great confidence in the truth of the fundamental principle on which our reasonings concerning the vibrations of elastic media have been based, the same degree of confidence is by no means extended to those adventitious suppositions which have been introduced for the sake of simplifying the analysis. (N.F. 284)

That he appeared on the defensive is hinted at by the inclusion of a footnote to the effect that his equations could be obtained by other (stated) means, but that he had used his chosen method because it gave rise to a particular equation which he would use afterwards (N.F. 286).

Green's output was the product of his middle age, from origins a lesser man would not have been able to overcome. What might have been had he lived into old age is a saddening thought. As it is, he was a self-taught genius who provided one of the bridges between the earlier generations of Laplace and Lagrange and the later generations of Thomson and Maxwell.

Appendix II

Mathematical Papers of George Green

1. An Essay on the Application of Mathematical Analysis to the Theories of Electricity and Magnetism. By George Green, Nottingham. Printed for the Author by T. Wheelhouse, Nottingham. 1828. (Quarto, vii + 72 pages.)
2. Mathematical Investigations concerning the Laws of the Equilibrium of Fluids analogous to the Electric Fluid, with other similar Researches. By George Green, Esq., Communicated by SirEdwardFfrenchBromhead,Bart.,M.A.,F.R.S.L.andE.(Cambridge Philosophical Society, read 12 November 1832, printed in the *Transactions* 1833. Quatro, 63 pages.) Vol. III, Part I.
3. On the Determination of the Exterior and Interior Attractions of Ellipsoids of Variable Densities. By George Green, Esq., Caius College. (Cambridge Philosophical Society, read 6 May 1833, printed in the *Transactions* 1835. Quarto, 35 pages.) Vol. III, Part III.
4. Researches on the Vibration of Pendulums in Fluid Media. By George Green, Esq., Communicated by Sir Edward Ffrench Bromhead, Bart., M.A., F.R.S.S. Lond. and Ed. (Royal Society of Edinburgh, read 16 December 1833, printed in the *Transactions* 1836. Quarto, 9 pages.) Vol. III, Part I.
5. On the Motion of Waves in a Variable Canal of Small Width and Depth. By George Green, Esq., BA, of Caius College. (Cambridge Philosophical Society, read 15 May 1837, printed in the *Transactions* 1838. Quarto, 6 pages.) Vol. VI, Part IV.
6. On the Reflexion and Refraction of Sound. By George Green, Esq., BA, of Caius College, Cambridge. (Cambridge Philosophical Society, read 11 December 1837, printed in the *Transactions* 1838. Quarto, 11 pages.) Vol. VI, Part III.

7. On the Laws of the Reflexion and Refraction of Light at the common Surface of two non-crystallized Media. By George Green, Esq., BA, of Caius College. (Cambridge Philosophical Society, read 11 December 1837, printed in the *Transactions* 1838. Quarto, 24 pages.) Vol. VII, Part I.
8. Note on the Motion of Waves in Canals. By George Green, Esq., BA, of Caius College. (Cambridge Philosophical Society, read 18 February 1839, printed in the *Transactions* 1839. Quarto, 9 pages.) Vol. VII, Part I.
9. Supplement to a Memoir on the Reflexion and Refraction of Light. By George Green, Esq., BA, of Caius College. (Cambridge Philosophical Society, read 6 May 1389, printed in the *Transactions* 1939. Quarto, 8 pages.) Vol. VII, Part I.
10. On the Propagation of Light in Crystallized Media. By George Green, BA, Fellow of Caius College. (Cambridge Philosophical Society, read 20 May 1839, printed in the *Transactions* 1839. Quarto, 20 pages.) Vol. VII, Part II.

Appendix IIIa

Account by William Tomlin, Esq.

Memoir of George Green Esq.

Late of Caius and Gonville College, Cambridge, and of Sneinton in the Cy. of Nottingham by William Tomlin, Nottingham 10th April 1845.

George Green Esq., B.A., and late Fellow of Gonville and Caius College, Cambridge, was born on the 14th July 1793, the only son of George and Sarah Green of Nottingham and Sneinton an adjacent village. He lived with his parents to the termination of their lives and duly rendered assistance to his Father in the prosecution of this businesses which was firstly a Baker at Nottingham and afterwards a Miller at Sneinton: in both cases being conducted on their own premises. But these assistances were irksome to the son who at a very early age and with in youth a frail constitution pursued with undeviating constancy the same as in his more mature years an intense application to Mathematics or whatever other acquirements might become accessary thereto. His schoolmasters soon perceiving this strong inclination for as well as profound knowledge in the mathematics and which far transcended their own, relinquished the direction of his studies and in consequence his literary acquirements were not properly promoted: in this respect, he had when contemplating the probability of going to the University to pay some attention in his more mature years.

Several kind and respected friends were anxious that he should adopt an University education at Cambridge several years before that circumstance actually took place, and they also assisted him in making known, and in the dedication of his first publication.

In the early part of the year 1829 his father who then only remained of his parents deceased at Sneinton, leaving him alone with the home and business on his hands; soon after which he disposed of the latter in order to be more at liberty to prosecute his mathematical studies to which he had now to add the Classics, and about the year 1836 he was duly entered on the books and became a member of Caius and Gonville College Cambridge. In the regular course he tried for University honours where to the disappointment of his friends he was only fourth wrangler, which circumstance was understood by them to arise from the easiness of the subjects given to the competitors and which is proved by the number of wranglers of that year being the greatest ever known. In November 1839 he was elected fellow of this college where he remained until the spring of the year 1840 when he returned indisposed after enjoying many years of excellent health to Sneinton, Alas! with the opinion that he should never recover from his illness and which became verified in little more than a year's time by his decease on the 31st May 1841. His remains were deposited in the same tomb with his parents in Sneinton Church Yard.

Mr. Green's published works are . . .

(the list as given by Mr. Tomlin is incomplete and slightly inaccurate: corrections and completions are made in the handwriting of William Thomson [Lord Kelvin]). (H.G.G. p. 582)

William Tomlin Esq. to Dr. J.C. Williams.

Nottingham April 9th 1845.

Dear Sir,

I well recollect the grateful feeling which my cousin and brother in law the late Mr. George Green entertained for the kindness shown to him by Sir E. ff. Bromhead and with this impression I think that I ought to accord with his wish and yours, by giving you some account of my above named relative, at the same time requesting your excuse for the imperfection of my narrative which you have herewith.

From Dr. Sir
Yours truly
William Tomlin

Dr. J.C. Williams
Nottingham

Dr. J.C. Williams to Sir E. Ffrench Bromhead

April 10th 1845

Dear Sir

I have just recived the enclosed letters which I forward to you – Mr. Tomlin married Mr. Green's sister, and he is a very respected man living on his Property in this town – I knew Mr. Green very well, and so far can vouch for the accuracy of Mr. Tomlin's statements – I remain Dear Sir Edward

Yours truly
J.C. Williams

Sir E. ff Bromhead, Bart.

Dr J.C. Williams was successively physician to General Hospital, Dispensary and Lunatic Asylum. He was a member of Bromley House Subscription Library, founder Trustee of Nottingham High School 1835, founder Trustee of Nottingham Mechanics Institution 1837, and Life Governor of Nottingham General Hospital 1844–1856. He died 21 July 1856 as a result of being thrown from his carriage near Lenton Toll-Bar on the Nottingham–Wollaton Turnpike [See *Nottingham Journal*, July 25, 1856]. Buried St Peter's Nottingham. (G.G.Cat. p. 10)

Appendix IIIb

Account by Sir E. Ffrench Bromhead

[From Sir E. Ffrench Bromhead to the Revd. J.J. Smith, Tutor of Gonville and Caius College.]

Lincoln
March 24 – 1845

My Dear Sir,

My acquaintance with the late Mr. Green was quite casual. I met with a subscription-list for his first matheml: publication, and added my name as Country Gentlemen often do by way of encouraging every attempt at provincial literature. When the work reached me it was obvious from the limited number of subscribers and their names that the publication must be a complete failure and dead born. On meeting with a person who knew him, I sent him word that I should be glad to see him and he visited me accordingly at Thurlby –

We of course had a great deal of scientific conversation, and he told me that he found so little sympathy with his pursuits, that he had finally relinquished them. Against this I warmly and successfully remonstrated (as i [*sic*] also have done in the like case of another distinguished Mathematician) explaining to him that such a publication as his not being elementary or systematic could not stand alone even in London and that against the Provincial press there was moreover a regularly organized conspiracy.

He mentioned to me some original views which he had, and I persuaded him to work them into the form of a memoir, which I undertook to lay for him before the Royal Societies of London or Edinburgh or our Society at Cambridge by which he would be saved from much trouble and expense and would find his views fairly brought before the scientific public and estimated at their value. We also conversed about Cambridge and on another occasion when he inclined to go there. I recommended my own college of Gonville and Caius at which with good conduct and a high degree he might be certain of a fellowship – I also gave him letters of introduction to some of the most distinguished characters of the University that he might keep his object steadily in view under some awe of their names and look upwards, not of course with any view of trespassing on the social distinctions of our University, in my time much more marked than at present, but that he might venture to ask advice under any emergency.

So much for my knowledge of poor Green, but I have written to a gentleman at Nottingham, who perhaps may supply further particulars and you will in the meantime

believe me
very truly yours
E.F. Bromhead

This correspondence is in the George Green Archive, Manuscripts Department, Nottingham University. GG Cat 4A 5, & 6.

Appendix IVa Green Family Tree

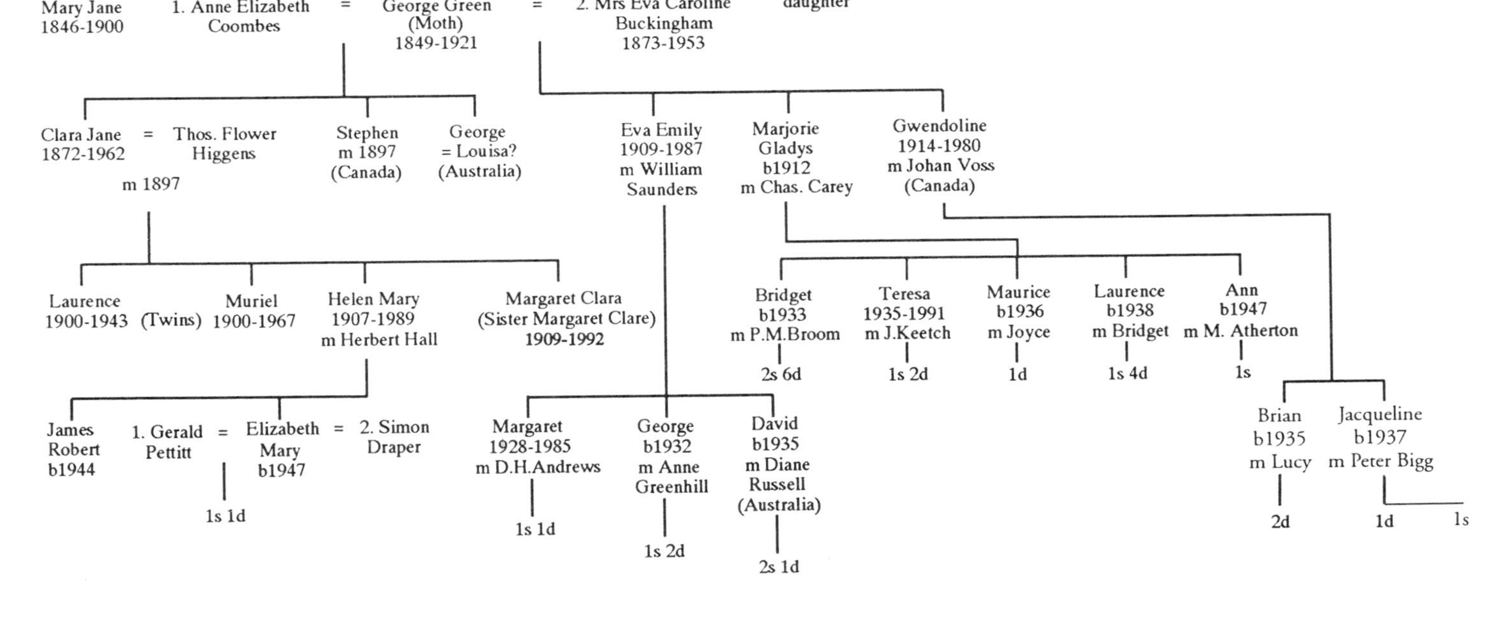
Mary Jane
1846-1900
1. Anne Elizabeth
Coombes
=
George Green
(Moth)
1849-1921
=
2. Mrs Eva Caroline
Buckingham
1873-1953
daughter
Clara Jane
1872-1962
=
Thos. Flower
Higgens
m 1897
Stephen
m 1897
(Canada)
George
= Louisa?
(Australia)
Eva Emily
1909-1987
m William
Saunders
Marjorie
Gladys
b1912
m Chas. Carey
Gwendoline
1914-1980
m Johan Voss
(Canada)
Laurence
1900-1943
(Twins)
Muriel
1900-1967
Helen Mary
1907-1989
m Herbert Hall
Margaret Clara
(Sister Margaret Clare)
1909-1992
Bridget
b1933
m P.M.Broom
2s 6d
Teresa
1935-1991
m J.Keetch
1s 2d
Maurice
b1936
m Joyce
1d
Laurence
b1938
m Bridget
1s 4d
Ann
b1947
m M. Atherton
1s
James
Robert
b1944
1. Gerald
Pettitt
=
Elizabeth
Mary
b1947
=
2. Simon
Draper
1s 1d
Margaret
1928-1985
m D.H.Andrews
1s 1d
George
b1932
m Anne
Greenhill
1s 2d
David
b1935
m Diane
Russell
(Australia)
2s 1d
Brian
b1935
m Lucy
2d
Jacqueline
b1937
m Peter Bigg
1d
1s

Appendix IVb Butler Family Tree

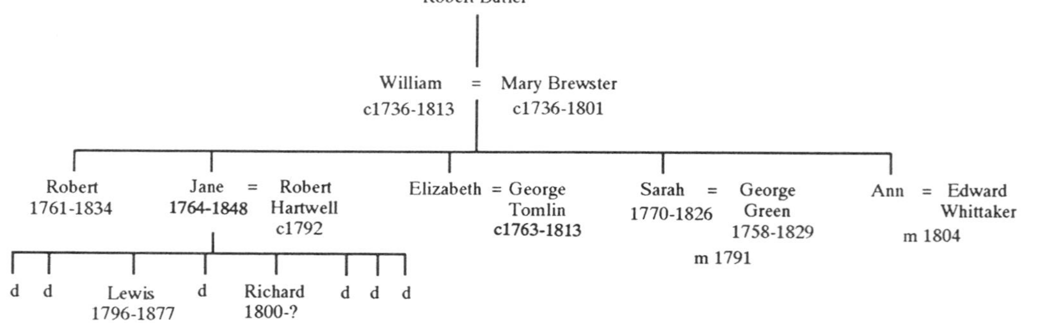

Appendix IVc Smith Family Tree

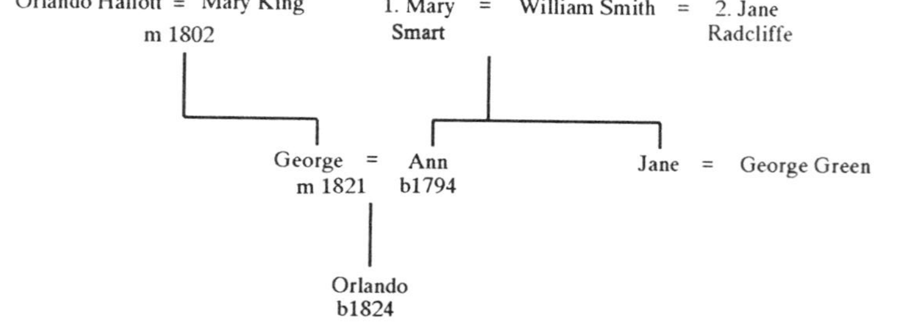

Appendix IVd Tomlin Family Tree

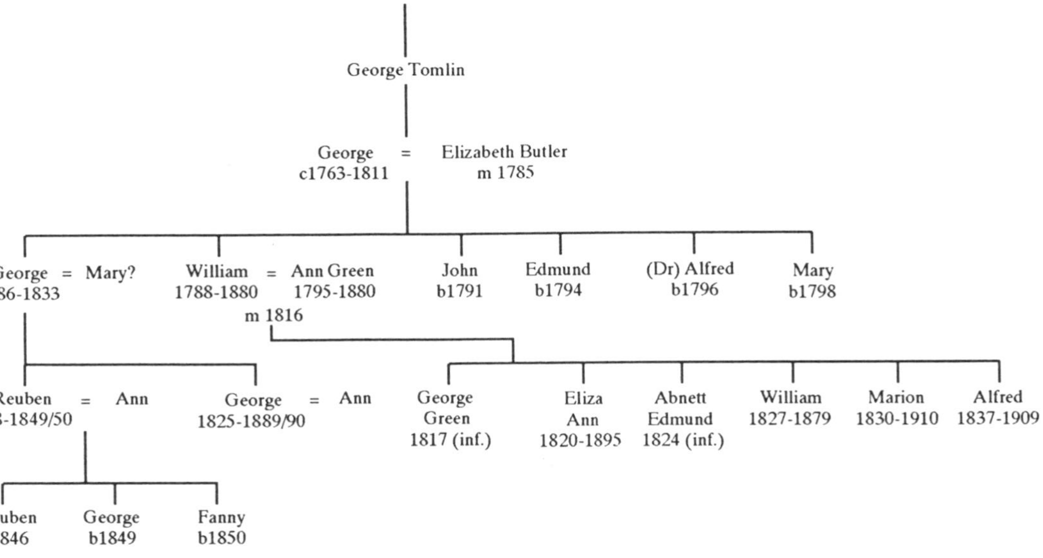

Appendix Va Time Chart of British Mathematicians and Men of Science
Mentioned in the Text

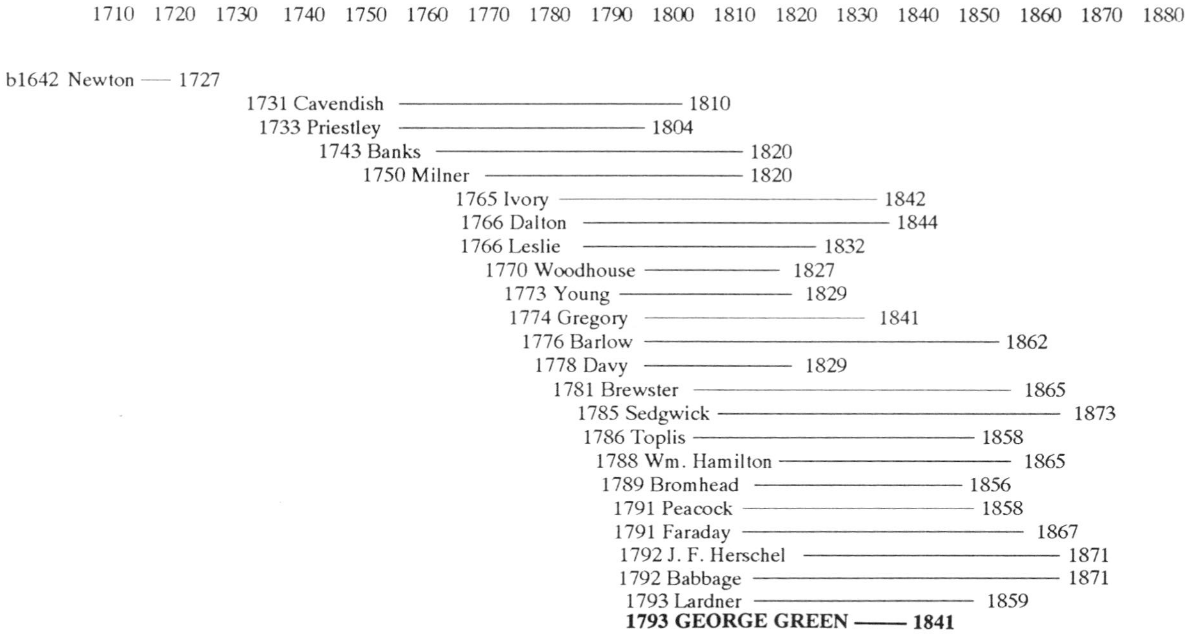

1710 1720 1730 1740 1750 1760 1770 1780 1790 1800 1810 1820 1830 1840 1850 1860 1870 1880

b1642 Newton — 1727
1731 Cavendish — 1810
1733 Priestley — 1804
1743 Banks — 1820
1750 Milner — 1820
1765 Ivory — 1842
1766 Dalton — 1844
1766 Leslie — 1832
1770 Woodhouse — 1827
1773 Young — 1829
1774 Gregory — 1841
1776 Barlow — 1862
1778 Davy — 1829
1781 Brewster — 1865
1785 Sedgwick — 1873
1786 Toplis — 1858
1788 Wm. Hamilton — 1865
1789 Bromhead — 1856
1791 Peacock — 1858
1791 Faraday — 1867
1792 J. F. Herschel — 1871
1792 Babbage — 1871
1793 Lardner — 1859
1793 GEORGE GREEN — 1841

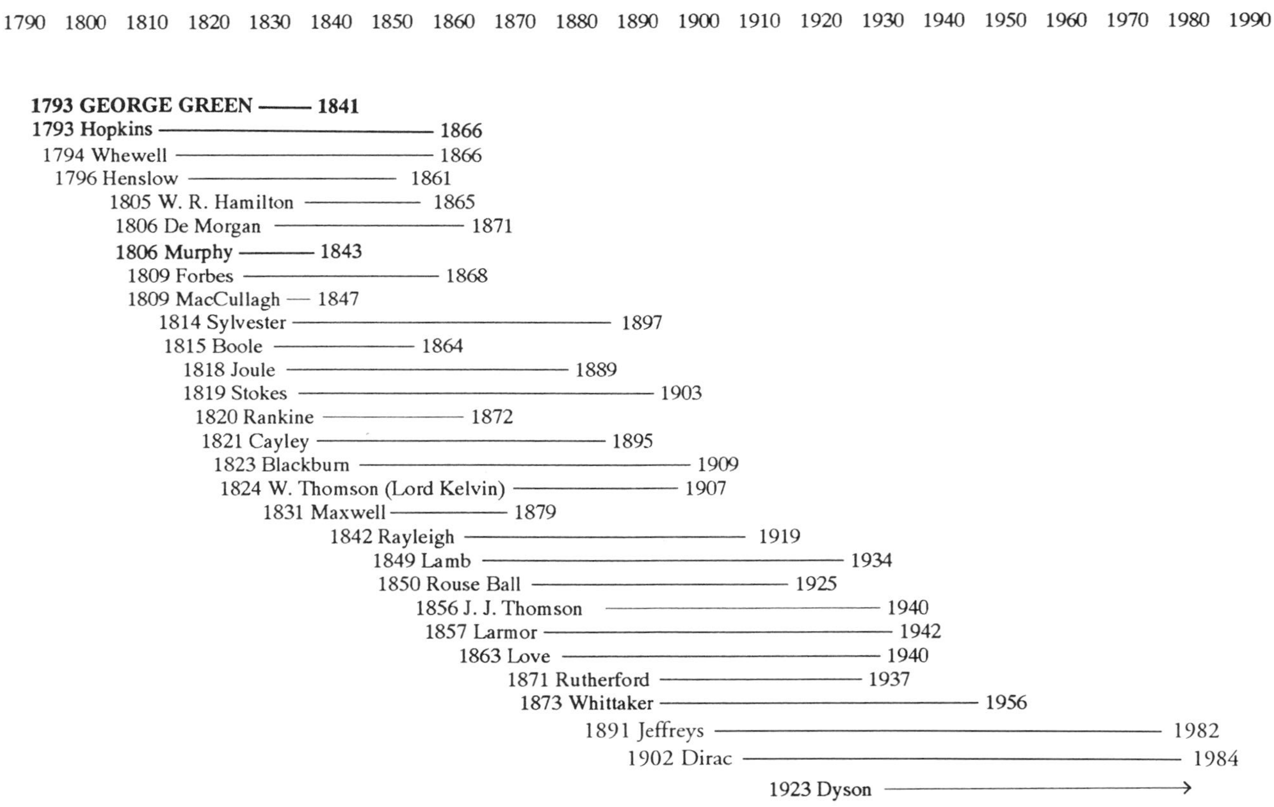

1790 1800 1810 1820 1830 1840 1850 1860 1870 1880 1890 1900 1910 1920 1930 1940 1950 1960 1970 1980 1990

1793 GEORGE GREEN —— 1841
1793 Hopkins —— 1866
1794 Whewell —— 1866
1796 Henslow —— 1861
1805 W. R. Hamilton —— 1865
1806 De Morgan —— 1871
1806 Murphy —— 1843
1809 Forbes —— 1868
1809 MacCullagh — 1847
1814 Sylvester —— 1897
1815 Boole —— 1864
1818 Joule —— 1889
1819 Stokes —— 1903
1820 Rankine —— 1872
1821 Cayley —— 1895
1823 Blackburn —— 1909
1824 W. Thomson (Lord Kelvin) —— 1907
1831 Maxwell —— 1879
1842 Rayleigh —— 1919
1849 Lamb —— 1934
1850 Rouse Ball —— 1925
1856 J. J. Thomson —— 1940
1857 Larmor —— 1942
1863 Love —— 1940
1871 Rutherford —— 1937
1873 Whittaker —— 1956
1891 Jeffreys —— 1982
1902 Dirac —— 1984
1923 Dyson ——→

Appendix Vb Time Chart of Other Mathematicians and Men of Science Mentioned in the Text

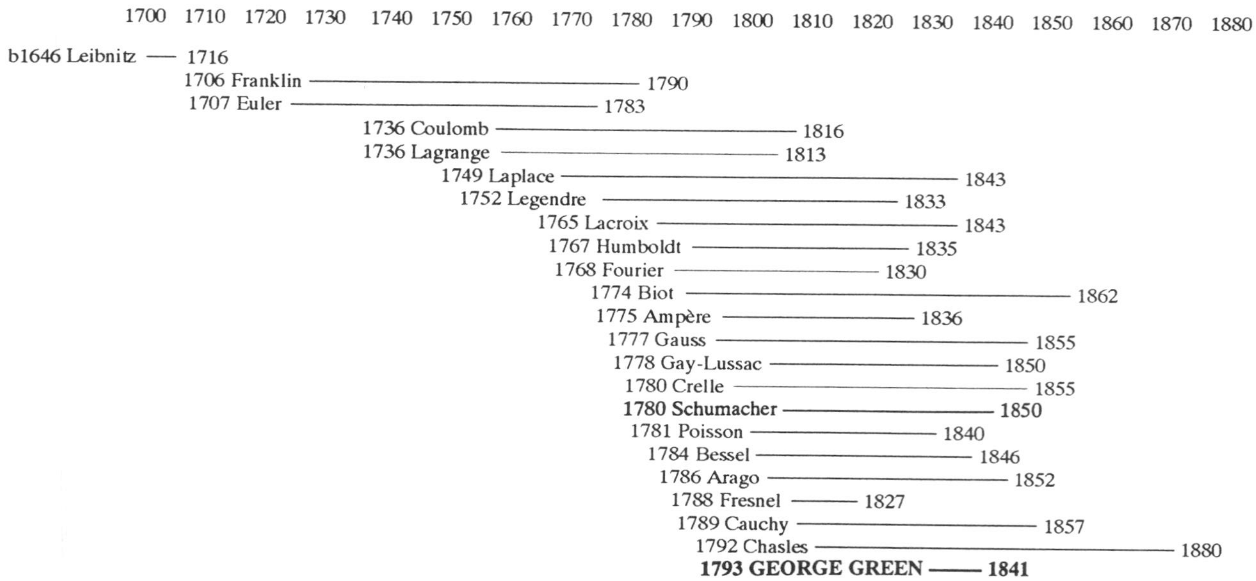

1700 1710 1720 1730 1740 1750 1760 1770 1780 1790 1800 1810 1820 1830 1840 1850 1860 1870 1880

b1646 Leibnitz — 1716
1706 Franklin — 1790
1707 Euler — 1783
1736 Coulomb — 1816
1736 Lagrange — 1813
1749 Laplace — 1843
1752 Legendre — 1833
1765 Lacroix — 1843
1767 Humboldt — 1835
1768 Fourier — 1830
1774 Biot — 1862
1775 Ampère — 1836
1777 Gauss — 1855
1778 Gay-Lussac — 1850
1780 Crelle — 1855
1780 Schumacher — 1850
1781 Poisson — 1840
1784 Bessel — 1846
1786 Arago — 1852
1788 Fresnel — 1827
1789 Cauchy — 1857
1792 Chasles — 1880
1793 GEORGE GREEN — 1841

1790 1800 1810 1820 1830 1840 1850 1860 1870 1880 1890 1900 1910 1920 1930 1940 1950 1960 1970 1980 1990

1793 GEORGE GREEN —— 1841
1798 Neumann —— 1895
1803 Sturm —— 1855
1804 Jacobi —— 1851
1805 Dirichlet —— 1859
1809 Liouville —— 1882
1810 Regnault —— 1878
1826 Riemann —— 1866
1854 Brillouin —— 1948
1875 de Broglie M. —— **1960**
1879 Einstein —— 1955
1885 Bohr —— 1962
1892 de Broglie L. —— 1960
1894 Kramers —— 1952
1898 Wentzel —— 1978
1906 Tomonaga —— 1979
1918 Feynman —— 1988
1918 Schwinger —— 1994
1921 Matsubara ——→

Appendix VIa

The Greening of Quantum Field Theory: George and I

Professor Julian Schwinger

The young theoretical physicists of a generation or two earlier subscribed to the belief that if you haven't done something important by age 30, you never will. Obviously, they were unfamiliar with the history of George Green, the Miller of Nottingham.

Born, as we all know, exactly two centuries ago, he received, from age 8, only a few terms of formal education. Thus, he was self-educated in mathematics and physics, when in 1828, at age 35, he published, by subscription, his first and most important work: *An Essay on the Application of Mathematical Analysis to the Theories of Electricity and Magnetism*. The Essay was dedicated to a noble patron of the 'Sciences and Literature', the Duke of Newcastle. Green sent his own copy to the Duke. I do not know if it was acknowledged. Indeed, as Albert Einstein is cited as effectively saying, during his 1930 visit to Nottingham, Green, in writing the Essay, was years ahead of his time.

There are those who cannot accept that someone, of modest social status and limited formal education, could produce formidable feats of intellect. There is the familiar example of William Shakespeare of Stratford-on-Avon. It took almost a century and a half to surface, and yet another century to strongly promote, the idea that Will of Stratford could not possibly be the source of the plays and the sonnets which had to have been written by Francis Bacon. Or was it the Earl of Rutland? Or perhaps it was William, the sixth Earl of Derby? The most recent pretender is Edward de Vir, Seventeenth Earl of Oxford, notwithstanding the fact that he had been dead for twelve years when Will was put to rest.

I have always been surprised that no one has suggested an analogous conspiracy to explain the remarkable mathematical feats of the Miller of Nottingham. So I invented one.

Descended from one of the lines of the Earl of Nottingham was the branch of the earls of Effindham, which was separated from the Howards in 1731. The fourth holder of the title died in 1816, with apparently no claimant. In that year, George Green, age 23, could well have reached the maturity that led, 12 years later, to the publication of the Essay. And what of the remarkable fact that, in the same year that the earldom was revived, 1837, George Green graduated fourth wrangler at Cambridge University?

The conspiracy at which I hint darkly is one in which I believe quite as much as I think Edward de Vir is the real Shakespeare.

I consider myself to be largely self-educated. A major source of information came from my family's possession of the *Encyclopaedia Brittanica*, Eleventh Edition. I recently became curious to know what I might have, and probably did, learn about George Green, some 65 years before.

There is no article detailing the life of George Green. There are, however, four brief references that indicate the wide range of Green's interests.

First, in the article 'Electricity,' as a footnote to the description of Lord Kelvin's work, is this:

> In this connexion the work of George Green (1793–1841) must not be forgotten. Green's *Essay on the Application of Mathematical Analysis to the Theories of Electricity and Magnetism*, published in 1828, contains the first exposition of the theory of potential. An important theory contained in it is known as Green's Theorem, and is of great value.

It was, of course, Lord Kelvin, or rather William Thomson, who rescued Green's work from total obscurity.

Then, in the article 'Hydromechanics,' after several applications of Green's transformation, which is to say, the theorem, there appears, under the heading *The Motion of a Solid through a Liquid:*

> The ellipsoid was the shape first worked out, by George Green, in his research on the vibration of a pendulum in a fluid medium (1833).

On to the article 'Light' under the heading *Mechanical Models of the Electromagnetic Medium.* After some negative remarks about Fresnel, one reads:

> Thus, George Green, who was the first to apply the theory of elasticity in an unobjectionable manner . . .

This is the content of *On the Laws of Reflexion and Refraction of Light* (1837).

Finally, the paper *On the Propagation of Light in Crystallized Media* (1839) appears in the Brittanica article 'Wave' as follows:

> The theory of waves diverging from a center in an unlimited crystalline medium has been investigated with a view to optical theory by G. Green.

The word 'propagation' is a signal to us that, in little more than 10 years, George Green had significantly widened his physical framework. From the static three-dimensional Green function that appears in potential theory, he had arrived at the concept of a dynamical, four-dimensional Green function. It would be invaluable a century later.

To continue the saga of George Green and me, my next step was to trace the influences of George Green on my own works. Here I spent no time over ancient documents. I went directly to a known source: THE WAR.

I presume that in Britain, unlike the United States, *the war* has a unique connotation. Apart from a brief sojourn in Chicago, to see if I wanted to help develop The Bomb—I didn't—I spent the war years helping to develop microwave radar. In the earlier hands of the British, that activity, famous for its role in winning the Battle of Britain, had begun with electromagnetic radio waves of high frequency, to be followed by very high frequency, which led to very high frequency indeed.

Through those years in Cambridge (Massachusetts, that is), I gave a series of lectures on microwave propagation. A small percentage of them is preserved in a slim volume entitled *Discontinuities in Waveguides.* The word *propagation* will have alerted you to the presence of George Green. Indeed, on pages 10 and 18 of an introduction there are applications of two different forms of Green's identity.

Then, on the first page of Chapter 1, there is Green's function, symbolized by G. In the subsequent 138 pages the references to Green in name or symbol are more than 200 in number.

As the war in Europe was winding down, the experts in high power microwaves began to think of those electric fields as potential electron accelerators. I took a hand in that and devised the microtron which relies on the properties of relativistic energy. I have never seen one, but I have been told that it works. More important and more familiar is the synchrotron.

Here, I was mainly interested in the properties of the radiation emitted by an accelerated relativistic electron. I used the four-dimensionally invariant proper time formulation of action. It included the electromagnetic self-action of the charge, which is to say that it employed a four-dimensionally covariant Green's function. I was only interested in the resistive part, describing the flow of energy from the mechanical system into radiation, but I could not help noticing that the mechanical mass had an invariant electromagnetic mass added to it, thereby producing the physical mass of an electron. I had always been told that such a union was not possible. The simple lesson? To arrive at covariant results, use a covariant formulation and maintain covariance throughout.

Quantum field theory, or more precisely, quantum electrodynamics, was forced from childhood into adolescence by the experimental results announced at Shelter Island early in June, 1947. The relativistic theory of the electron created by Dirac in 1928 was wrong. Not very wrong, but measurably so.

A few days later, I left on a honeymoon tour across the United States. Not until September did I begin to work on the obvious hypothesis that electrodynamic effects were responsible for the experimental deviations, one on the magnetic moment of the electron, the other on the energy spectrum of the hydrogen atom.

Although a covariant method was in order, I felt I could make up time with the then more familiar non-covariant methods of the day. By the end of November I had the results. The predicted shift in magnetic moment agreed with experiment. As for the energy shift in hydrogen, one ran into an expected problem.

Consider the electromagnetic momentum associated with a charge moving at constant speed. The ratio of that momentum to the speed is a mass—an electromagnetic mass. It differs from the electromagnetic mass inferred from the electromagnetic energy. Analogously, the magnetic dipole moment inferred for an electron moving in an electric field is wrong. Replacing it by the correct dipole moment leads to an energy level displacement that was correct in 1947, and remains correct today at that level of accuracy as governed by the fine structure constant.

I described all this at the January 1948 meeting of the American Physical Society, after which Richard Feynman stood up and announced that he had a relativistic method. Well, so did I, but I also had the numbers. Indeed, several months later, at the opening of the Pocono Conference, he ran over to me, shook my hand, and said 'Congratulations, Professor! You got it right', which left me somewhat bewildered. It turned out he had completed his own calculation of the additional magnetic moment. Later we compared notes and found much in common.

Unfortunately, one of the things we shared was an incorrect treatment of low energy photons. Nothing fundamental was involved; it was a matter of technique in making a transition between two different gauges. But, as in American politics these days, the less important the subject, the louder the noise. When that lapse was set right, the result of 1947 was regained. Incidentally, even Lord Rayleigh once made a mistake. That's one reason for its being called the Rayleigh–Jeans law.

To keep to the main thrust of the talk—the evolution of Green's function in the quantum mechanical realm—I move on to 1950, and a paper entitled *On Gauge Invariance and Vacuum Polarization.*

This paper makes extensive use of Green's functions, in a proper-time context, to deal with a variety of problems: non-linearities of the electromagnetic field, the photon decay of a neutral meson, and a short, but not the shortest, derivation of the additional electron magnetic moment. The latter ends with the remark that 'The concepts employed here will be discussed at length in later publications'. I cannot believe I wrote that.

The first, rather brief, discussion of those concepts appeared in a pair of 1951 papers, entitled *On the Green's Functions of Quantized Fields.* One would not be wrong to trace the origin of today's lecture back 42 years to these brief notes. This is how paper I begins:

> The temporal development of quantized fields, in its particle aspect, is described by propagation functions,

> or Green's functions. The construction of these functions for coupled fields is usually considered from the viewpoint of perturbation theory. Although the latter may be resorted to for detailed calculations, it is desirable to avoid founding the formal theory of the Green's functions on the restricted basis provided by the assumption of expandability in powers of the coupling constants. These notes are a preliminary account of a general theory of Green's functions, in which the defining property is taken to be the representation of the fields of prescribed sources.

We employ a quantum dynamical principle for fields which has been described in the 1951 paper entitled *The Theory of Quantized Fields*. This (action) principle is a differential characterization of the function that produces a transformation from eigenvalues of a complete set of commuting operators on one space-like surface to eigenvalues of another set on a different surface.

In one example of a rigorous formulation, Green's function, for an electron-positron, obeys an inhomogeneous Dirac differential equation for an electromagnetic vector potential that is supplemented by a functional derivative with respect to the photon source, and the vector potential obeys a differential equation in which the photon source is supplemented by a vectorial part of the electron-positron Green's function. (It looks better than it sounds.) It is remarked that, in addition to such one-particle Green's functions, one can also have multiparticle Green's functions.

The second note begins with:

> In all the work of the preceding note there has been no explicit reference to the particular states on (the space-like surfaces) that enter the definitions of the Green's functions. This information must be contained in boundary conditions that supplement the differential equations. We shall determine these boundary conditions for the Green's functions associated with vacuum states on both (surfaces).

And then:

> We thus encounter Green's functions that obey the temporal analog of the boundary condition characteristic of a source radiating into space. In keeping with this analogy, such Green's functions can be derived from a retarded proper-time Green's function by a Fourier decomposition with respect to the mass.
>
> The text continues with the introduction of auxiliary quantities: the mass operator M that gives a non-local extension to the electron mass; a somewhat analogous photon polarization operator P; and Γ, the non-local extension of the coupling between the electromagnetic field and the fields of the charged particles. Then, in the context of two-particle Green's functions, there is the interaction operator I.
>
> The various operators that enter in the Green's function equations M,P,Γ,I, can be constructed by successive approximation. Perturbation theory, as applied in this manner, must not be confused with the expansion of the Green's functions in powers of the charge. The latter procedure is restricted to the treatment of scattering problems.

Then one reads:

> It is necessary to recognize, however, that the mass operator, for example, can be largely represented in its effect by an alteration in the mass constant and by a scale change of the Green's function. Similarly, the major effect of the polarization operator is to multiply the photon Green's function by a factor, which everywhere appears associated with the charge. It is only after these renormalizations have been performed that we deal with wave equations that involve the empirical mass and charge, and are thus of immediate physical applicability.

In the period 1951–1952, two colleagues of mine at Harvard, and I, wrote a series of papers under the title *Electrodynamic Displacements of Atomic Energy Levels*. The third paper, which does not carry my name, is subtitled *The Hyperfine Structure of Positronium*. I quote a few lines:

> The discussion of the bound states of the electron-positron system is based upon a rigorous functional differential equation for the Green's function of that system.

And:

> Theory and experiment are in agreement.

As for the rest of the 1950s, I focus on two highlights. First: Although it could have appeared any time after 1951, it was 1958 when I published *The Euclidian Structure of Relativistic Field Theory*. Here is how it begins:

> The nature of physical experience is largely conditioned by the topology of space-time, with its indefinite Lorentz metric. It is somewhat remarkable, then, to find that a detailed correspondence can be established between relativistic quantum field theory and a mathematical image based on a four-dimensional Euclidian manifold. The objects that convey this correspondence are the Green's functions of quantum field theory, which contain all possible physical information. The Green's functions can be defined as vacuum-state expectation values of time-ordered field products.

I well recall the reception this received, running the gamut from 'It's wrong' to 'It's trivial'. It is neither.

Second (highlight): Another Harvard colleague and I spent quite some time evolving the techniques before we published a 1959 paper entitled *Theory of Many-Particle Systems*. It was intended to bring the full power of quantum field theory to bear on the problems encountered in solid state physics, for example. That required the extension of vacuum Green's functions, which refer to absolute zero temperature, into those for finite temperature. This is accomplished by a change of boundary conditions, which become statements of periodicity, or anti-periodicity, for the respective BE or FD statistics, in respect to an imaginary time displacement.

As an off shoot of this paper, I published in 1960, *Field Theory of Unstable Particles*. Here is how it begins:

> Some attention has been directed recently to the field theoretic description of unstable particles. Since this question is conceived as a basic problem for field theory, the responses have been some special device or definition, which need not do justice to the physical situation. If, however, one regards the description of unstable particles to be fully contained in the framework of the general theory of Green's function, it is only necessary to emphasize the relevant structure of these functions. That is the purpose of this note. What is essentially the same question, the propagation of excitations in many-particle systems where stable or long-lived 'particles' can occur under exceptional circumstances, has already been discussed along these lines.

One might be forgiven for assuming that this saga of George and me effectively ended with this paper. But that was 1/3 century ago!

To set the stage for what actually happened, I remind you that operator field theory is an extrapolation of ordinary quantum mechanics, with its finite number of degrees of freedom, to a continuum labelled by the spatial coordinates. The use of such space-time dependent variables presumes the availability, in principle, of unlimited amounts of momentum and energy. It is, therefore, a hypothesis about all possible phenomena of that type, the vast majority of which lies far outside the realm of accessible physics. In honor of a failed economic policy, I call such procedures: trickle-down theory.

In the real world of physics, progress comes from tentative excursions beyond the established framework of experiment and theory—the grass roots—indeed, the Green grass roots. What is sought here, in contrast with the speculative approach of trickle-down theory, is a phenomenological theory—a coherent account of the phenomena that is anabatic (from anabassis: going up).

The challenge was to reconstruct quantum field theory, without operator fields. The source concept was introduced in 1951 as a mathematical device—it was a source of fields. It took 15 years to appreciate that, with a finite, rather than an unlimited, supply of energy available, it made better sense to use the more physical—if idealized—concept of a particle source. Indeed, during that time period one had become accustomed to the fact that to study a particle of high energy physics, one had to create it. And, the act of detection involved the annihilation of that particle.

This idea first appeared in an article, entitled *Particles and Sources*, which recorded a lecture of the 1966 *Tokyo Summer Lectures in Theoretical Physics*. The preface begins with:

> It is proposed that the phenomenological theory of particles be based on the source concept, which is abstracted from the physical possibility of creating or annihilating any particle in a suitable collision. The source representation displays both the momentum (energy) and the space-time characteristics of particle behaviour.

Then, in the introduction, one reads:

> Any particle can be created in a collision, given suitable partners, before and after the impact to supply the appropriate values of the spin and other quantum numbers, together with enough energy to exceed the mass threshold. In identifying new particles it is basic experimental principle that the specific reaction is not otherwise relevant. Then, let us abstract from the physical presence of the additional particles involved in creating a given one (this is the vacuum) and consider them simply as the source of the physical properties that are carried by the created particle. The ability to give some localization in space and time to a creation act may be represented by a corresponding coordinate dependence of a mathematical source function, $S(x)$. The effectiveness of the source in supplying energy and momentum may be described by another mathematical source function, $S(p)$. The complementarity of these source aspects can be given its customary quantum interpretation: $S(p)$ is the four-dimensional Fourier transformation of $S(x)$.

The basic physical act begins with the creation of a particle by a source, followed by the propagation (aha!) of that particle between the neighborhoods of emission and detection, and is closed by the source annihilation of the particle. Relativistic requirements largely constrain the structure of the propagation function—Green's function.

We now have a situation in which Green's function is not a secondary quantity, implied by a more fundamental aspect of the theory, but rather, is a primary part of the foundation of that theory. Of course fields, initially inferred as derivative concepts, are of great importance, as witnessed by the title I gave to the set of books I began to write in 1968: *Particles, Sources, and Fields*.

The quantum electrodynamics that began to emerge in 1947 still bothers some people because of the divergences that appear prior to renormalization. That objection is removed in the phenomenological source theory where there are no divergences, and no renormalization.

As another example of such clarification I cite a 1975 paper entitled *Casimir Effect in Source Theory*. The abstract reads:

> The theory of the Casimir effect, including its temperature dependence is rederived by source theory methods, which do not employ the concept of (divergent) zero point energy. What source theory does have is a photon Green's function, which changes in response to the change of boundary conditions, as one conducting sheet is pushed into the proximity of another one.

A few years later, I, and two colleagues at the University of California (UCLA), who had joined me from Harvard with their new doctorates, extended this treatment to dielectric bodies where forces of attraction also appear.

Having said this, I can move up to the present day and the fascinating phenomenon of coherent sonoluminescence.

It has only recently been discovered that a single air bubble in water can be stabilized by an acoustical field. And, that the bubble emits pulses of light, including ultraviolet light, in synchronism with the sonic frequency.

During the phase of negative acoustical pressure the bubble expands. That is followed by a contraction which, as Lord Rayleigh already recognized in his 1917 study of cavitation, turns into run away collapse. The recent measurements find speeds in excess of Mach 1 in air.

Then the collapse abruptly slows, and a blast of photons is emitted. In due time, the expansion slowly begins, and it all repeats, and repeats.

When confronted with a new phenomenon, everyone tends to see in it something that is already familiar. So, when told about this new aspect of sonoluminescence, I immediately said 'It's the Casimir effect'! Not the static Casimir effect, of course, but the dynamical one of accelerated dielectric bodies. I have had no occasion to change my mind.

I can imagine a member of this audience thinking: 'That's nice, but what is the role of George Green in this'?

Looking in at the center of the water container, one sees a steady blue light. A photo-multiplier tube registers the succession of pulses, each containing a substantial number of photons, which can be an incomplete count because, deep in the ultraviolet, water becomes opaque.

A quantum mechanical description seeks the probabilities of emitting various numbers of photons, all of which probabilities are referred to the basic probability, that for emitting no photons. The latter probability dips below one—in some analogy with synchrotron radiation—because of the self-action carried by the electromagnetic field, as described by Green's function. And that function must obey the requirements imposed by an accelerated surface discontinuity, with water, the dielectric material, on one side, and a dielectric vacuum, air, on the other side. Carrying out that program is—as one television advertiser puts it—job one. Very fascinating, indeed.

So ends our rapid journey through 200 years. What, finally, shall we say about George Green? Why, that he is, in a manner of speaking, alive, well and living among us.

Appendix VIb

Homage to George Green: How Physics Looked in the Nineteen-Forties

Professor Freeman Dyson

1. Concepts and Tools

I am delighted and honoured to be chosen to speak here at the birthplace of Nottingham's great mathematician George Green, together with my old friend and benefactor Julian Schwinger. Julian Schwinger is my benefactor because he allowed me forty-five years ago to become famous by publishing his ideas. His generosity at that time was just as remarkable as the equal generosity of Richard Feynman whose death we are now mourning. Let me now say simply, thank you Julian, for without your help I would not be standing here today.

I am delighted to be talking about George Green because his life and work exemplify two of my favourite themes, two themes that are still today highly relevant both to the progress and to the public understanding of science. The first theme is the perennial and usually unsuccessful struggle to keep the doors of the temple of science open to amateurs and outsiders. George Green was a prime example of an amateur and outsider, without any official academic credentials, beating the insiders at their own game. He was lucky to have lived in the early nineteenth century rather than in the late twentieth century. He was, in spite of his social and educational deficiencies, allowed to enter the temple, and his achievements were recognized by the insiders. If George Green were living today, since science has become professionalized and the PhD has become a necessary ticket for admission to the temple, he would have encountered much more formidable barriers to his ambitions. The insiders are now defending their turf against outsiders with bureaucratic weapons unknown in the eighteen-thirties.

The second theme that George Green's work exemplifies is the historical fact that scientific revolutions are more often driven by new tools than by new concepts. Thomas Kuhn in his famous book, *The Structure of Scientific Revolutions*, talked almost exclusively about concepts and hardly at all about tools. His idea of a scientific revolution is based on a single example, the revolution in theoretical physics that occurred in the nineteen-twenties with the advent of quantum mechanics. This was a prime example of a concept-driven revolution. Kuhn's book was so brilliantly written that it became an instant classic. It misled a whole generation of students and historians of science into believing that all scientific revolutions are concept-driven. The concept-driven revolutions are the ones that attract the most attention and have the greatest impact on the public awareness of science, but in fact they are comparatively rare. In the last five hundred years we have had six major concept-driven revolutions, associated with the names of Copernicus, Newton, Darwin, Maxwell, Einstein and Freud, besides the quantum-mechanical revolution that Kuhn took as his model. During the same period there have been about twenty tool-driven revolutions, not so impressive to the general public but of equal importance to the progress of science. I will not attempt to make a complete list of tool-driven revolutions. Two prime examples are the Galilean revolution resulting from the use of the telescope in astronomy, and the Crick-Watson revolution resulting from the use of X-ray diffraction to determine the structure of big molecules in biology. The effect of a concept-driven revolution is to explain old things in new ways. The effect of a tool-driven revolution is to discover new things that have to be explained. In physics there has been a preponderance of tool-driven revolutions. We have been more successful in discovering new things than in explaining old ones. George Green's discovery, the Green's function, is a mathematical tool rather than a physical concept. It did not give the world a new theory of electricity and magnetism or a new picture of physical reality. It gave the world a new bag of mathematical tricks, useful for exploring the consequence of theories and for predicting the existence of new phenomena that experimenters could search for. The Green's function was a tool of discovery, like the telescope and the microscope, but aimed at mathematical models and theories instead of being aimed at the sky and the microbe.

The invention of the Green's function brought about a tool-driven revolution in mathematical physics, similar in character to the more famous tool-driven revolution caused by the invention of electronic computers a century and a half later. Both the Green's function and the computer increased the power of physical theories, particularly in the fields of electromagnetism, acoustics and hydrodynamics. The Green's function and the computer are prime examples of intellectual tools. They are tools for clear thinking. They helped us to think more clearly by enabling us to calculate more precisely.

But I did not come here today to preach to you about the evils of the PhD system and the importance of tools in science, two subjects that I have frequently talked about on other occasions. I was invited here by Professor Challis (and I quote) 'to give a lecture on the events associated with the first introduction of Green's functions to a quantum mechanical treatment of electrodynamics'. So I am here as a historical monument, like the windmill on the hill nearby, as a witness of events long past in which both I and George Green played a role. I am supposed to tell you what happened in 1948 when the words 'Green's function', which had been part of the accepted language of classical electrodynamics and fluid mechanics for a hundred years, suddenly began to be spoken by quantum theorists and became part of the fashionable jargon in the new field of quantum electrodynamics. I decided not to do precisely what Professor Challis asked. If I should try to describe the events of 1948 in detail, I would have to fill several blackboards with equations, and my story would be intelligible only to those members of the audience who already know it. I decided to interpret Professor Challis's request broadly. Instead of giving you a lecture on quantum electrodynamics, I will talk in a more general way about the physics of the nineteen-forties. I will try to give you an impression of the scientific communities in Europe and the United States as they then existed. I hope to explain how it happened that my own modest activities as a messenger transmitting knowledge from Europe to America turned out to be useful.

2. Classical and Quantum Green's Functions

To understand what happened in the nineteen-forties we must begin with some historical background. There are two kinds of physics, classical physics beginning with Galileo and Newton in the seventeenth century, and quantum physics beginning with Planck and Bohr in the twentieth century. Classical physics describes big things such as rocks and planets. Quantum physics describes small things such as atoms and electrons. Next, cutting across the division of physics into classical and quantum, there is a division of physical objects into discrete and continuous. A rock is a discrete object. A flowing liquid or a magnetic field is a continuous object. Discrete objects are described by a finite set of numbers specifying their positions and velocities. The physics of discrete objects is called mechanics. Continuous objects are described by fields specifying their distribution and movement in space and time. The physics of continuous objects is called field theory. We have then four varieties of physical theories: classical mechanics, classical field theory, quantum mechanics and quantum field theory.

A highly compressed account of the history of theoretical physics goes like this. Physics is a drama in six acts. Act one, classical physics of discrete objects, was worked out by Galileo and Newton. Act two, a hundred years later, classical physics of continuous objects, was worked out by Euler and Coulomb and Oersted. Euler did hydrodynamics, Coulomb did electrostatics and Oersted did magnetism. So it happened that, at the beginning of the nineteenth century, the classical field theories of hydrodynamics and electrostatics and magnetism were well established. Act three: George Green in 1828 revolutionized classical field theory by introducing his new tool, the Green's function, which described directly the causal relationship between the behaviour of a field at any two points in space and time. The Green's function measures the local response of the field at a given point at a later time to a local disturbance of the field at another given point at an earlier time. Green used the Green's function to clarify in a fundamental way the causal relationships between electric and magnetic fields. Helmholtz subsequently used Green's functions to clarify in an equally fundamental way the causal relationships between pressure and velocity in acoustics. Act four: Heisenberg and Schrödinger worked out the quantum physics of discrete objects, describing the behaviour of atoms and electrons with the theory that became known as quantum mechanics. Act five: Fermi and Heisenberg and Dirac invented quantum field theory to describe the quantum physics of continuous objects. The quantum field theory that described electricity and magnetism was called quantum electrodynamics. But quantum field theory did not work well as a practical tool. It was unreliable and tended to give absurd answers to simple questions. You asked a quantum field theory the question, 'What is the mass of an electron'? and the answer came back, 'Infinity'. That was not very helpful. As a result of these well-publicized absurdities, the majority of practical physicists, especially in America, wrote off quantum field theory as useless and probably wrong. So at the end of act five in the nineteen-thirties we had physics divided into two disconnected parts, the classical field theories which worked beautifully in the classical domain, and the quantum mechanics of particles which worked beautifully in the quantum domain. There was no connection between the two domains. Green's functions were a convenient working tool in the classical domain, but there were no Green's functions in the quantum domain. The quantum field theories, which should have been the link between the two domains, were discredited and generally believed to be useless. That was the situation at the beginning of the nineteen-forties. Act six: the resurrection of quantum field theories and the introduction of quantum Green's functions at the end of the nineteen-forties. Act six is the main subject of this lecture.

In order to set the stage for act six, I looked at the four books out of which I learned physics as a student to see how often the name of George Green appears in them. The four books, which I still have as treasured possessions on my shelves, are *Theoretical Physics* by George Joos, written in 1932, *The Principles of Quantum Mechanics* by Paul Dirac, written in 1930, *The Quantum Theory of Radiation* by Walter Heitler, written in 1935, and *Quanthentheorie der Wellenfelder* by Gregor Wentzel, written in 1942. I will have more to say in a moment about the Wentzel book. All four books are classics, full of beautiful writing and clear thinking. I still refer to them frequently as sources of useful information, not only as historical relics. It turns out that George Green is mentioned only twice, by Joos and by Dirac, in both cases in connection with Green's Theorem. Green's Theorem is one of Green's major contributions to science, establishing an exact relation between the sources and fluxes of two fields. It relates the sources of two fields inside any region of space to the fluxes of the same fields through the surface bounding the region. The Theorem is applied by Joos to a problem in classical electrostatics; by Dirac to a problem in quantum scattering of a particle. But Green's more important discovery, the Green's function, is not mentioned by name in any of the books. If you look closely at the Heitler and the Wentzel books, you will see that Green's functions are lurking on many of their pages, but they are not labeled as such. The Green's functions appear mainly in equations and are called commutators or potentials when they are mentioned in the text. I only learned much later, after I became a professor and had to learn something of the history of physics in order to teach it, that these elegant and useful tools had been borrowed from George Green.

3. Nicholas Kemmer

I now begin my narrative of act six as I experienced it, first in England and then in America. In 1946 I came to Cambridge (the real Cambridge, not the American imitation) with the intention of learning modern physics. When I arrived there, I found that experimental physics was at a low ebb. The experimenters had been away during the war. In 1946 they were still struggling to get started on new enterprises which were to achieve huge success within a few years, the beginnings of the new sciences of radio-astronomy and molecular biology. I understood that Martin Ryle with his radio receivers and Max Perutz with his haemoglobin crystals were doing this exciting stuff, but the stuff they were doing was clearly not physics. If I had wanted to be in a place where world-class experimental physics was being done, I should have gone to Bristol. In Bristol, Cecil Powell and his team of scanners with their microscopes and photographic emulsions were developing the techniques which led them within two years to the discovery of the pi-meson. But I was by training a mathematician, a student of Besicovitch, who had taught me the fine art of combining geometrical with analytical reasoning. I enjoyed talking with experimenters, but my more urgent need was to talk to a competent mathematical physicist. I needed to find somebody in Cambridge who could tell me what the important unsolved problems in theoretical physics were, and how I might use my mathematical skills to solve them.

My first enormous stroke of luck was to find Nicholas Kemmer in Cambridge. He was the teacher I needed. He rapidly became a friend as well as a teacher, and I am happy to say that our friendship is still alive and well after forty-five years. Kemmer gave two courses of lectures in Cambridge, one on nuclear physics and one on quantum field theory. The quantum field theory course was a distillation of the wisdom of Europe, at that time still unknown in America. It happens that quantum field theory, a rigidly formal mathematical discipline, was invented in Europe and was for a long time more highly regarded in Europe than in America. In 1946 the only existing text-book on quantum field theory was the book, *Quantentheorie der Wellenfelder*, by Gregor Wentzel, written in Zurich and published in 1943 in Vienna in the middle of the war. Kemmer possessed a copy of Wentzel's book and allowed me to borrow it. It was at that time a treasure without price. I believe there were then only two copies in England. It was later reprinted in America and translated into English. But in 1946, few people in America knew of its existence and even fewer considered it important. Kemmer knew it was important. He not only lent it to me but also explained why it was important.

Kemmer belonged to the generation of scientists whose careers were maximally disrupted by the war. As a young man in 1938 he had published a theory of nuclear forces mediated by a symmetric triplet of meson fields, one positive, one negative and one neutral. The purpose of the symmetric triplet was to achieve equality of the neutron-proton and proton-proton forces. In 1938 not one of the three hypothetical mesons had yet been discovered. The symmetric meson theory was considered a wild speculation. Ten years later, all three mesons were found, and the theory was proved to be a correct description of a new symmetry of nature. It is in fact one of the most brilliant predictions in the history of physics, comparable in brilliance with Yukawa's original prediction of the existence of the meson. But Kemmer received little public acclaim when his theory was confirmed. To blow his own trumpet was not in his nature. In the meantime, he had spent most of the wartime years working on the Canadian atomic energy project at Chalk River. After the war, he was given a lecturing job in Cambridge with a heavy teaching load and an enormous number of undergraduates to supervise. I was appalled to see the long hours he had to spend supervising students every day. There was nobody else who supervised students as conscientiously as Kemmer, and so they came to him in huge numbers. He never complained, but he paid a heavy price for being such an excellent supervisor. He had no time left over to resume the research career so brilliantly begun ten years before. Kemmer and I were both living in Trinity College. He was a college lecturer and was treated by the college as a drudge, while I was a junior fellow with no duties and complete freedom to do whatever I liked. This was a monstrously unfair division of labour, but Kemmer seemed to accept it without any sign of resentment. He was as generous in spending time with me as he was with his students. He always had time to advise me, to explain the difficult points in Wentzel's book, and to share with me his vision of quantum field theory as the key to a consistent mathematical description of nature. He was, and is, the most unselfish scientist I ever knew.

During my year at Cambridge I decided to go to America and make a fresh start there. In spite of my friendship with Kemmer, I found Cambridge depressing. I wanted to be in a place where I would be involved in an active group of young people doing research. By chance I met Sir Geoffrey Taylor, who had a little hand-made wind-tunnel in a cellar under the Cavendish Laboratory and did classic experiments on turbulence in the classic Cavendish string-and-sealing-wax style. He was also the world's greatest expert on blast-waves, and had been at Los Alamos during the war to make sure that the bombs were exploded at the correct height to achieve the maximum blast damage. I told him I was planning to go to America and asked where I should go. He said at once, 'Oh, you should go to Cornell and work with Bethe. That is where all the brightest people from Los Alamos went when the war was over'. The conversation was over in one minute. At that time I hardly knew that Cornell existed, but I took Sir Geoffrey's advice, and a year later I was a student of Hans Bethe and a friend of Richard Feynman. That was my second enormous piece of luck.

4. Physics in 1948

I have said enough about my personal history. Let me now turn to a description of the state of physics as I found it during my first year in America, the year 1948. My personal memories of the events of that year are hopelessly unreliable after forty-five years have gone by. A much more reliable and informative view of the state of American physics is obtained by looking at Volume 73 of the *Physical Review*, the leading American physics journal which was started at

Cornell University in 1893. In the nineteen-forties the journal came out twice every month. Volume 73 contains the issues from January to June 1948. It covers all areas of physics, experimental and theoretical, atomic and nuclear and astronomical, quantum and classical.

Looking through Volume 73 today, I see a great number of familiar faces belonging to old friends. Almost every paper in it is interesting, and many of them are memorable. In 1948 the issues of the journal were thin enough so that we could read them from cover to cover. Many of us did. Nowadays, of course, the journal is fragmented into six parts, each of which is so fat that nobody even attempts to read it. In 1948 it was possible to read the whole journal and obtain an overview of everything that American physicists were doing.

The paper that impressed me most strongly in 1948 and still impresses me today is entitled *Relaxation Effects in Nuclear Magnetic Resonance Absorption*, a monumental piece of work, thirty-four pages long, by Bloembergen, Purcell and Pound, three Harvard physicists. All three of them are still going strong and two of them are still at Harvard. I worked through all the details of this massive work with intense pleasure. Nuclear magnetic resonance means the tickling of nuclear magnets inside a piece of solid or liquid material by alternating magnetic fields applied from the outside. Nuclear magnetic resonance was in 1948 a recently discovered phenomenon. It is now, forty-five years later, the basis of the medical technique known as MRI, magnetic resonance imaging, which enables doctors to obtain clear pictures of brain tumours and other soft tissue abnormalities in their patients. The paper of Bloembergen, Purcell and Pound addresses the question, What are the effects of the environment in which the nuclei are embedded on the detailed behaviour of their magnetic resonance? The paper reports a comprehensive series of experiments together with an equally comprehensive theoretical analysis. It is one of the finest examples of the American style of physics, with experiment and theory working together as inexorably as a steam-roller and squashing a problem flat. The paper provided a fundamental understanding of the various ways in which the physical and chemical properties of the environment are linked with the shape of the nuclear resonance. It demonstrated that the nuclear resonance could be made into a powerful new tool for exploring the properties of matter. It provided the essential foundation of knowledge on which the development of MRI as a tool of medical diagnosis could be built thirty years later.

In the same volume of the *Physical Review* are many other wonderful papers on the most diverse subjects: Alpher, Bethe and Gamow on the origin of chemical elements. Gleb Wataghin on the formation of chemical elements inside stars. Edward Teller on the change of physical constants. Lewis, Oppenheimer and Wouthuysen on the multiple production of mesons. Foley and Kusch on the experimental discovery of the anomalous magnetic moment of the electron. Julian Schwinger on the theoretical explanation of the anomalous moment. Paul Dirac on the quantum theory of localizable dynamical systems. This was a vintage year for historic papers. I mention these seven just to give you the flavour of what physicists were doing in the early post-war years. The Alpher–Bethe–Gamow paper proposed that the chemical elements were formed by successive capture of neutrons on protons during the initial expansion of the universe from a hot dense beginning. Bethe had nothing to do with the writing of the paper. He allowed his name to be put on it to fill the gap between Alpher and Gamow. This joke, which was of course Gamow's idea, made the paper famous. Meanwhile, the paper of Wataghin, which proposed that the elements were formed in neutron stars, or more precisely in the process of rapid expansion of neutron stars into interstellar space, received much less attention. Wataghin was then living in Brazil and was not widely known. Unfortunately, it took us many years to collect the evidence which proved that, at least for the great majority of the elements, Alpher–Bethe–Gamow were wrong and Wataghin was right. I do not have time today to discuss the other papers on my list. Volume 73 contained several hundred papers, almost all of them worth reading.

One fact which I found remarkable in 1948 and still find remarkable today is that, among the hundreds of papers in Volume 73, the paper of Dirac is the only one concerned with quantum field theory. Dirac was a voice from another world. The vast majority of the papers are like the paper of Bloembergen, Purcell and Pound, sticking close to experiments and using a minimum of theory. The American scientific tradition was strongly empirical. Theory was regarded as a necessary evil, needed for the correct understanding of experiments but not valued for its own sake. Quantum field theory had been invented and elaborated in Europe. It was a sophisticated mathematical construction, motivated more by considerations of mathematical beauty than by success in explaining experiments. The majority of American physicists had not taken the trouble to learn it. They considered it, as Samuel Johnson considered Italian opera, an exotic and irrational entertainment. The paper of Dirac, although it was published in America, found few readers. It was read by the small community of experts in general relativity who were themselves isolated from the mainstream of American physics.

Thus, it happened that I arrived at Cornell as a student, and found myself, thanks to Nicholas Kemmer, the only person in the whole university who knew about quantum field theory. The great Hans Bethe and the brilliant Richard Feynman taught me a tremendous lot about many areas of physics, but when we were dealing with quantum field theory I had to be the teacher and they had to be the students. Bethe and Feynman had been doing physics successfully for many years without the help of quantum field theory, and so they were not eager to learn it. It was my luck that I arrived with this gift from Europe just at the moment when the new precise experiments of Lamb and others on the fine details of atomic energy levels required quantum field theory for their correct interpretation. When I used quantum field theory to calculate an actual number, the Lamb shift separating the energy levels of two of the states in a hydrogen atom with a spinless electron, Bethe was impressed. He said it was the first time he had seen quantum field theory do anything useful. For Bethe, formal mathematical machinery was pointless unless it could be used for calculating numbers. For him, and for almost all American theorists at that time, calculating numbers was the object of the game. Since the little gift that I brought from Europe to America could be used for calculating numbers, and the numbers could be checked by experiment, the gift received a friendly reception.

5. Quantum Field Theory and Green's Functions

Julian Schwinger had, of course, known all about quantum field theory long before. But he shared the American view that it was a mathematical extravagance, better avoided unless it should turn out to be useful. In 1948 he understood that it could be useful. He used it for the fine details of atomic physics revealed by the experiments of Lamb and Retherford, Foley and Kusch at Columbia. But he used it grudgingly. In his publications he preferred not to speak explicitly about quantum field theory. Instead, he spoke about Green's functions. It turned out that the Green's functions which Schwinger talked about and the quantum field theory that Kemmer talked about were fundamentally the same thing. In Schwinger's papers I could recognize some of my old friends, functions that I had seen before in Wentzel's book. This was one of the ways that Green's functions came to occupy a central place in the particle physics of the nineteen-fifties.

The second way that Green's functions emerged in particle physics was through the work of Richard Feynman at Cornell. Feynman had never been interested in quantum field theory. He had his own private way of doing calculations in particle physics. His way was based on things that he called 'Propagators', which were probability amplitudes for particles to propagate themselves from one space-time point to another. He calculated the probability of physical processes by adding up the propagators. He had rules for calculating the propagators. Each propagator was represented graphically by a collection of diagrams. Each diagram gave a pictorial view of particles moving along straight lines and colliding with one another at points where the straight lines met. When I learned this technique of drawing diagrams and calculating propagators from Feynman, I found it completely baffling, because it always gave the right answers but did not seem to be based on any solid mathematical foundation. Feynman called his way of calculating physical processes 'the space-time approach', because his diagrams represented events as occurring at particular places and at particular times. The propagators described sequences of events in space-time. It later turned out that Feynman's propagators were identical with Green's functions. Feynman had been talking the language of Green's functions all his life without knowing it.

The third way that Green's functions appeared in the particle physics of the nineteen-forties was in the work of Sin-Itiro Tomonaga, who had developed a new and elegant version of relativistic quantum field theory. His work was done in the complete isolation of war-time Japan, and was published in Japanese in 1943. The rest of the world became aware of it only in the spring of 1948, when an English translation of it arrived in Princeton, sent by Hideki Yukawa to Robert Oppenheimer. Tomonaga was a physicist in the European tradition, having worked as a student with Heisenberg at Leipzig before the war. For him, in contrast to Schwinger and Feynman, quantum field theory was a familiar and natural language in which to think about particle physics. Tomonaga and Dirac were on the same wavelength. In the paper of Dirac which I mentioned earlier, published in the 1948 *Physical Review*, Dirac mentions Tomonaga in the text and in a footnote, but does not refer either to Schwinger or to Feynman.

After the war, Tomonaga's students in Japan had been applying his ideas to calculate the properties of atoms and electrons with high accuracy, and were reaching the same results as Schwinger and Feynman. When Tomonaga's papers began to arrive in America, I was delighted to see that he was speaking the language of quantum field theory that I had learned from Kemmer. It did not take us long to put all the various ingredients of the pudding together. When the pudding was cooked, all three versions of the new theory of atoms and electrons turned out to be different ways of expressing the same basic ideas. The basic idea of all three ways was to calculate Green's functions for all atomic processes that could be directly observed. Green's functions appeared as the essential link between the methods of Schwinger and Feynman, and Tomonaga's relativistic quantum field theory provided the firm mathematical foundation for all three versions of quantum electrodynamics.

6. The Later History

The history of physics did not end in 1950. In this talk I will only sketch very briefly some highlights of the later history. One of the major early advances beyond Tomonaga, Schwinger and Feynman was made by Rudolf Peierls in Birmingham in 1951 and published in a paper, *The Commutation Laws of Relativistic Field Theory*, in the Proceedings of the Royal Society. Peierls is like Kemmer, an exceptionally unselfish person as well as a brilliant physicist. When I returned from America to England in 1949, he welcomed me as a member of his department in Birmingham and gave me all the privileges that he never asked for himself. He had a heavy teaching load; mine was minimal. He and his wife Genia were responsible for a big household and four children; I was a guest in their home. The idea was that he would take care of all the mundane chores while I would have freedom and leisure to make important discoveries in physics. Of course, things did not work out the way they were planned. During my two years in Birmingham, the most brilliant discovery made in the Peierls department was the general commutation law of relativistic field theories. The discovery was his, not mine. I remember the surprise and delight when he told me about his discovery in the garden of his house on Carpenter Road. It gave us for the first time a deep and general understanding of the connection between Green's functions and commutation relations between fields. It also clarified the meaning of the correspondence principle which connects classical with quantum field theories. It put Green's functions where they belong, in the logical foundation of classical and quantum physics.

During the nineteen-fifties, the new methods of calculation, using Green's functions to describe the behaviour of quantum fields, were successfully applied to a variety of problems in electrodynamics. Experiments and theory were pushed to higher and higher levels of accuracy. When we began these calculations we had hoped to find a clear discrepancy between theory and experiment. A discrepancy would reveal some fundamental information about the many known and unknown particles and interactions that the theory of quantum electrodynamics does not take into account. Since quantum electrodynamics does not pretend to be a complete theory of everything, it must at some level of accuracy disagree with experiment. But all attempts to find a discrepancy ended in disappointment. As each quantity was measured and calculated to more and more places of decimals, the measured and calculated numbers remained obstinately equal. Quantum electrodynamics turned out to be a more accurate description of nature than anybody in the nineteen-forties had imagined possible. It now agrees with experiment to ten or eleven places of decimals. We are left with an unsolved mystery to explain. How could all the important fields and particles that lie outside the scope of quantum electrodynamics have conspired to hide their influence on the processes that lie inside? To solve the mystery, even more accurate experiments and calculations will be required.

After the Green's function method has been successfully applied to quantum electrodynamics, the next big step was to apply the same method to the description of many-electron systems in the physics of condensed matter. I began the application to condensed matter physics in 1956 with a study of spin-waves in ferromagnets. I found that all the Green's function tricks that had worked so well in quantum electrodynamics worked even better in the theory of spin-waves. The spin-wave is the simplest propagating mode of disturbance in the condensed assemblage of electrons inside a ferromagnet, just as the photon is the simplest propagating mode in the electromagnetic field in free space. I was able to calculate the scattering of one spin-wave by another, using the same tricks that Feynman had used in quantum electrodynamics to calculate the scattering of light by light. Meanwhile, the Green's function method was applied systematically by Bogolyubov and other people to a whole range of problems in condensed matter physics. The main novelty that arose when we moved into condensed matter physics was the appearance of temperature as an additional variable. In quantum electrodynamics we had considered atoms and electrons in free space, in an environment with zero temperature. In condensed matter the temperature can never be zero, and many of the most interesting questions concern the effect of temperature on the properties of the system. The appropriate tools for analyzing condensed-matter properties are therefore thermal Green's functions, Green's functions describing matter in thermal equilibrium at a given temperature. A beautiful thing happens when you make the transition from ordinary Green's functions to thermal Green's functions. To make the transition, all you have to do is replace the real time by a complex number whose real part is the time and whose imaginary part is the temperature. Thus thermal Green's functions are just as easy to calculate as ordinary Green's functions. To put in the temperature, you simply give the time an imaginary component. This is mathematical magic which I will not attempt to explain. Green's functions make such magic possible. That is one of the sources of their power and their beauty.

Soon after thermal Green's functions were invented, they were applied to solve the outstanding unsolved problem of condensed matter physics, the problem of superconductivity. They allowed Bardeen, Cooper and Schrieffer to understand superconductivity as an effect of a particular thermal Green's function expressing long-range phase-coherence between pairs of electrons. The Bardeen–Cooper–Schrieffer theory of 1957 explained satisfactorily all the observed features of superconductors as they were then known. The only thing the theory did not do was to give any hint of the existence of the high-temperature superconductors that were discovered unexpectedly thirty years later.

In the nineteen-sixties, after Green's functions had become established as the standard working tools of theoretical analysis in condensed matter physics, the wheel of fashion in particle physics continued to turn. For a decade, quantum field theory and Green's functions in particle physics were unfashionable. The prevailing view was that quantum field theory had failed in the domain of strong interactions, and that only phenomenological models of strong interaction processes could be trusted. Then, in the nineteen-seventies, the wheel of fashion turned once more. Quantum field theory was back in the limelight with two enormous successes, the Weinberg-Salam unified theory of electromagnetic and weak interactions, and the gauge theory of strong interactions now known as quantum chromodynamics. Green's functions were once again the working tools of calculation, both in particle physics and in condensed matter physics. And so they have remained up to the present day.

In the nineteen-eighties, quantum field theory moved off into new directions, to lattice gauge theories in one direction and to superstring theories in another. The Wilson Loop is the reincarnation of a Green's function in lattice gauge theory. I have no doubt that there is a corresponding reincarnation of Green's functions in superstring theory, but I stop my narrative at this point in order not to expose my ignorance further. As we move into the nineteen-nineties, Green's functions are alive and well, still going strong, ready to help us again as soon as the wheel of fashion turns once more and the next new theory of everything emerges. That is not the end of the story, but it is the end of my talk.

Notes

List of Abbreviations Used in the Notes

B.C.S.	R.M. Bowley, L.J. Challis and F.W. Sheard, 'George Green, His Achievements and Place in Science', in *George Green, Miller, Snienton*, Castle Museum, Nottingham, 1976.
B.L.	British Library.
B.H.L.	Bromley House Library.
D.N.B.	*Dictionary of National Biography.*
D.P.	David Phillips, 'George Green: His Academic Career', in *George Green, Miller, Snienton*, Castle Museum, Nottingham, 1976.
D.S.B.	*Dictionary of Scientific Biography.*
F.W.J.	Freda Wilkins-Jones, 'George Green, His Family and Background', in *George Green, Miller, Snienton*, Castle Museum, Nottingham, 1976.
G.G.Cat.	*George Green: A Catalogue of Books and Manuscripts Associated with Green*, Department of Manuscripts, Nottingham University Library.
G.C.C.	Gonville and Caius College Cambridge
H.G.G.	H. Gwynedd Green, 'A Biography of George Green', in M. Ashley Montague (ed.), *Studies and Essays in the History of Science and Learning*, (New York, 1945), pp. 545–94.
N.C.L.	Nottingham County Library (Local Studies Department).
N.F.	N. Ferrers (ed.), *Mathematical Papers of George Green* (Cambridge, 1871), republished by the Chelsea Publishing Company, New York, 1971. Page references for Green's papers are taken from this edition.

N.R.O.	Nottinghamshire Record Office.
N.U.L.	Nottingham University Library.
R.B.N.	Record of the Borough of Nottingham.
R.S.Cat.	Royal Society Catalogue of Scientific Papers.
S.P.T.	S.P. Thompson, *Life of Lord Kelvin*, London: Macmillan, 1910.
T.C.P.S.	*Transactions of the Cambridge Philosophical Society.*
T.R.S.	*Transactions of the Royal Society.*
T.Th.S.	*Transactions of the Thoroton Society of Nottinghamshire*, ed. Adrian Henstock, Nottinghamshire Record Office, and Hazel Salisbury, printed for the Society by Derry & Sons, Nottingham.

Chapter 1 Family Background

1. F.W.J. p. 15.
2. I am grateful to Miss Kathleen Harmon for first drawing my attention to their existence. Some ten years ago, it was possible to decipher the inscriptions; the overgrowth and further weathering would now make that impossible.
3. Archdeacon's Registry of Marriage Bonds for Nottinghamshire.
4. N.R.O. M 1874. The Green farm area was first identified by Mrs Freda Wilkins-Jones.
5. I am grateful to Mr John Cupit for showing me round his farm, and giving me this information.
6. F.W.J. p. 18
7. Ibid.
8. *The Illustrated Journeys of Celia Fiennes*, ed. Christopher Morris (Exeter: Webb & Bower, 1892), p. 86. 'Nottingham was Celia's favourite town which she used as a basis of comparison' (editor's footnote, p. 97).
9. F.W.J. p. 18, quoting Robert Sanders's *Complete English Traveller* (1772).
10. Archive Teaching Unit 2.
11. Archive Unit 1. See also Chambers, p. 6.
12. H.G.G. p. 570.
13. F.W.J. p. 44.
14. Ibid. p. 31.

15. Ibid. p. 22.
16. Sep. 1. (1800) A terrible riot in the town: the mob broke the bakers' windows; 2nd, they came to Mr. Pepper's house and went into all the villages around.
    Sep. 11. More rioting on account of high price of flour...
    Sep. 16. The 1st Dragoon Guards, 4 troops of them, came into town after 12 o'clock at night from Northampton to keep the mob quiet . . . (*Diary of Abigail Gawthern 1751–1810*, ed. Adrian Henstock, p. 83: Thoroton Society Record Series, vol. XXXIII)
17. N.R.O. CA 3990/1/1. See Wells, ch. 6, pp. 26–36.
18. F.W.J. p. 34, quoting from the *Nottingham Journal*, 14 February 1807.
19. Quoted by Mellors, p. 70. Colwick, a neighbouring village, still markets a cream cheese.
20. F.W.J. p. 39, quoting from the *Nottingham Evening Post*, 26 June 1923.
21. Tomlin's letter of 1845. G.G.Cat. 41 6 ii.
22. Chaucer, *Canterbury Tales*, transl. Neville Coghill (Penguin, 1954), p. 109.
23. I am grateful to Mr Tom Davies of Messrs Thompson, Alford, Lincolnshire, millwrights, and Mr David Bent, miller at Green's Mill, Sneinton, for their help in compiling this account.
24. F.W.J. p. 43.
25. Ibid. p. 44.
26. Ibid. p. 43.
27. Chambers, p. 27.
28. I am grateful to the late Mr Harold Wiseman for showing me round his house.
29. N.R.G. DD H158 DD CA 118.
30. Letter from Dr J.C. Williams to Sir Edward Bromhead, 1845. See Appendix IIIa, G.G.Cat. 4A 6 ii.
31. Mellors, p. 98. The clock was installed *c.* 1856.
32. F.W.J. p. 48. The Elizabeth Teague Charity was founded by Elizabeth Teague, who died in Hanover Square, London, in 1776 and left £100 for the poor of Sneinton, the income from which was to provide bread and coals on St Thomas's Day. Mellors, p. 32.
33. F.W.J. p. 48. H.G. Green gives Percy J. Cropper's Manuscript

on Sneinton as reference on these matters. This is in the Special Collections in the MSS. Department at Nottingham University, MS. 1468/3. Cropper cites George Green, churchwarden, as one of four who, in 1824, fixed the Poor Rate at one shilling in the pound 'on the value of land and houses in the parish of Snenton'. 'George Green Mill etc.' was rated at £2.4s.4d. – one of only three inhabitants in Old Sneinton rated above £2 per annum: MS. 1468/4.

34. As summarized by H.G. Green, p. 580.
35. N.R.O. CA 3989/1353. See F.W.J. p. 48. The declaration is signed 'George Green, Miller, Snienton'. An alternative, archaic form of spelling was 'Snenton'; the modern version is Sneinton.
36. F.W.J. p. 52.
37. Ibid.
38. Lace-dressing was a process in which the lace was soaked in a mixture containing glue and starch, then stretched on a frame to dry and stiffen.
39. J.V. Beckett, 'The Church of England and the Working Class in Nineteenth Century Nottingham', T.Th.S. XCII, 1988, p. 61.
40. H.G.G. pp. 575–6.
41. See below, Chapter 9, p. 140, and reference to the sampler 'marked' by Jane Green; also George Green Junior's declaration, Chapter 9, p. 125.
42. *Kelly's Directory*, 1854.

Chapter 2 George Green's Education

1. G.G.Cat. 4A 6.
2. N.R.O. M/373/25. F.W.J. p. 27.
3. Chapman, p. 38.
4. John Blackner writes in his *History of Nottingham* (1815): 'If the adoption of any science, scheme or institution, ever received the approbation of mankind more than any other, it is that of teaching the children of the indigent the rudiments of education on the Sabbath day.' (p. 127).
5. Chapman, p. 36.
6. Wardle, p. 38.

7. Chambers, p. 32. R.B.N. records the case of a mill girl accused of stealing, who said she earned 7 shillings a week plus 1*s.* 6*d.* for seaming at nights. Twenty shillings made one one pound (converted to 100 pence on metrication in 1972); a guinea was one pound and one shilling, for which there was a single coin.
8. Wardle, p. 18.
9. Blackner, pp. 122–6.
10. A.W. Thomas, p. 23.
11. 'Wrangler' is the term used in Cambridge to indicate graduates gaining first-class honours in mathematics. Names are now published alphabetically; in Green's time they were published in order of merit. To be 'Senior Wrangler' at the top of the list was an honour coveted both by the individual and by his college, equalled only by becoming a Smith's Prizeman (see below, Chapter 3, Note 6).
12. A.W. Thomas, p. 119.
13. Ibid., p. 120.
14. Gawthern (ed. Henstock), p. 26.
15. Ibid., p. 27.
16. Chapman, p. 160.
17. Richard Bonington (1768–1835), father of the painter Richard Parkes Bonington (1802–28).
18. Henry Kirke White (1785–1805). His work was taken up by Robert Southey, the Poet Laureate, and enjoyed a considerable vogue in the nineteenth century.
19. Record of the Family of Goodacre. B.H.L. G.G.
20. R. Goodacre, 'Prospectus'. N.C.L. L37.3. Standard.
21. Wadsworth, p. 63.
22. An electrical machine would have produced a high voltage when a handle was turned. A 'quadrant' was similar in principle to a protractor, shaped like a quarter- rather than a half-circle, and graduated for making angular measurements. An 'orrery' was a clockwork model of the planetary system, devised *c.* 1700, and named after the Earl of Orrery.
23. Briggs Collection, N.U.L. Special Collections.
24. Goodacre, *Essay on the Education of Youth*, p. 44–5.
25. Ibid., p. 24.
26. B.H.L. Record of the Family of Goodacre.

    Goodacre had constructed an original apparatus to illustrate the principles of astronomy, and launched on a career as a lecturer, giving public lectures on astronomy with practical illustrations. In 1821 he gave his first lecture in a Bradford theatre, assisted by his son William, and described the event in a letter to his elder son Robert, who was left in charge of the Academy. One passage is worth quoting as an example of his enthusiasm, his rhetorical style and his approach to a scientific subject:

> When I entered through the stage door a breathless silence instantly succeeded and for the first time in my life I felt totally deprived of utterance. . . . I quickly recovered my embarrassment and proceeded fluently without being able to perceive whether I was admired or not, but the most intense and fixed attention was visible through the whole assembly. At length when I had uttered the patriotic exclamations against the Popish Inquisition which followed the case of Galileo, a lively round of applause followed. These were three or four times repeated during my quoting Shakespeare and Lord Byron and again when we exposed the Globe which William did in very good style. As I turned it about I pointed out the cause of the alternate light and shade which is thrown over the distant audience in a pleasing way and ventured to wish we might live to see the day when the Continent of Africa might exhibit as much scientific and moral brightness as it did on my globe. This touched the string of sympathy and . . . applause followed. (Ibid.)

27. F.W.J. p. 42.

Chapter 3 Cambridge Interlude

1. Amos, p. 34.
2. I am indebted to Dr John Twigg, historian of Queens' College, Cambridge, for this information.
3. Article on Isaac Milner, D.N.B.
4. Milner, p. 298.
5. D.N.B. For a more detailed account of Isaac Milner, see Twigg, various references.
6. Robert Smith (1689–1768), graduate and Fellow of Trinity

College and Master from 1742, made a bequest to the University enabling prizes to be awarded to two Junior BAs 'who have made the greatest progress in mathematics and natural philosophy'. The *Gentleman's Magazine* of April 1827 noted that the value of the prizes was £25 each. In the course of time the status of Smith's Prizeman rivalled in prestige that of Senior Wrangler. Mary Milner quotes interesting accounts by Herschel, Peacock and others of their examination by Milner: pp. 524 ff.

7. I am grateful to Mr Iain Wright, Curator of the Old Library, Queens' College, for allowing me access to these Inventories and alerting me to Milner's interest in French mathematics.
8. Milner, p. 695.
9. First noted in Dubbey, p. 26. The article, published in the *Philosophical Magazine*, vol. XX, 1805, pp. 25–31, is entitled 'On the Decline of Mathematical Studies and the Sciences dependent on them'.
10. The first printing press was set up in Cambridge in 1521 by J. Siberch who, a few years later, also started selling books. In 1529 Cardinal Wolsey was petitioned by the University to 'have the privilege to buy books of foreign merchants, as in the Universities beyond the seas' ('Bookselling in Cambridge, and the Booksellers of Ancient Times', *The Library World*, vol. 33, 1930–31, pp. 49–56). By Milner's time, booksellers had proliferated: their trade was highly competitive and their stocks were extensive, as their catalogues show – for example Deighton's of 1819–20.

    Thus Cambridge booksellers would probably have attracted keen interest from Fellows and ex-students on visits to the town, especially those from provincial and country areas. Edward Bromhead writes to his friend Charles Babbage of one such visit on 1 October 1819: 'No french [*sic*] mathematics lying about at Deighton's. French books sell well, Garnier especially. A new edition of Woodhouse's Trigonometry with small changes. . . . Works on Divinity have increased four to one in all the shops' (B.L. Add. MSS. 37, 182, fo. 99).

    Deighton's 1819–20 catalogue does not list texts by French mathematicians, which would be ordered by individual request. The 1837 catalogue however (when Green was already at Cambridge), lists 21 titles by Biot, Lacroix,

Lagrange, Laplace, Legendre and Poisson (see also Feather, ch. 5, pp. 69–98, 123–4). Feather's book, though it deals with the growth of the provincial book trade in the eighteenth century, sets the scene for its continued development in the nineteenth.

11. Whittaker, p. 153.
12. Ball, p. 83.
13. J.M. Thomas, 'Michael Faraday and the Royal Institution'.
14. Grattan-Guinness, 1985, p. 85.
15. Whittaker, p. 153.
16. Grattan-Guinness, 1981, p. 96.
17. Reference has been made to entries in the D.S.B. for general information on French men of science. For a full discussion of the French contribution to the development of mathematical physics at this time, see Ball, pp. 7–27.
18. The Académie des Sciences was known as the Mathematical and Physical Class of the Institut de France during the revolutionary period 1795–1815. For an account of publications by the Académie and in scientific journals, see Grattan-Guinness, 1981, pp. 101–4.
19. Letter from Herschel to Babbage written from Slough, 27 July 1819, inscribed at the top: 'Corresponding Member of the Royal Academy of Göttingen'. (B.L. Add. MSS. 37, 182, fo. 152). When Sir William Herschel, who was born in Hanover, died in 1822, his sister Caroline, who had helped her brother with his astronomical observations, returned to Hanover, where she was visited by Gauss, Humboldt and other eminent men of science. Her nephew, John Frederick, visited her in 1824 and 1832. Herschel and Babbage went to Paris in 1819 and met Arago, Biot and Fourier, amongst others. They also went to Arcueil, where Bertollet, Laplace and scientific friends ran the Société d'Arcueil, which functioned as an influential scientific *salon* and published its own scientific papers. See Maurice Crosland, *The Society of Arcueil* (London: Heinemann, 1976).
20. Artz, pp. 87–162. See also Magliulo, *Les Grandes Ecoles* (Paris: Presses Universitaires de France, 1982).
21. '. . . and now the most distinguished science faculty in the world' (Artz, p. 255). For its development, see Grattan-Guinness, 1981, pp. 106–8.

22. This was not the case in Scottish universities, where lectures in natural philosophy included experiments and demonstrations. See William Thomson's experience, Chapter 10, p. 145 below.
23. For an account of the quarrel see Aiton, pp. 63–7.
24. For a mathematical account of the development of the calculus in France and England, see Grattan-Guinness, 1981, pp. 97–101; 1985, pp. 95–6.
25.

> Every one acquainted with the recent history of Cambridge studies will be ready to assign to him [Woodhouse] the high praise of having been the first to introduce mathematics to us in the form which the advancing researches of continental mathematics had given to the sciences. (William Whewell, *Free Motion of Points and on Universal Gravitation . . . The First Part of a New Edition of a Treatise on Dynamics* [Cambridge: Cambridge University Press, 1832], Preface, p. iv)

Woodhouse's books proved valuable to an able student like Babbage, but Whewell had reservations about the average undergraduate:

> But I do not think it has been found a convenient book for the Cambridge student, and it appears to me difficult to introduce any considerable portion of it into the usual course of reading. Within the last few years however the deficiency which existed in this respect has been admirably supplied by Professor Airy's 'Mathematical Tracts'.

George Biddell Airy, appointed to Newton's Chair of Mathematics in 1826 at the age of twenty-five, published in that year his *Mathematical Tracts*, which included sections on astronomy and the calculus. In the second edition (1831), he added a chapter 'On the Undulatory Theory of Light'. He later wrote on sound and magnetism. These books were 'Designed for the Use of Students in the University', and were still on sale nearly fifty years later. But Whewell considered continental mathematics too difficult: 'The admirable systematic treatises of Laplace and Lagrange . . . are not suited to the common student, and can never be familiarly consulted except by accomplished and persevering analysts.'

Whewell appears always to have been cautious. He was not sure Peacock would be appointed Examiner a second time after his temerity in introducing the continental notation into the 1817 examination paper, and he was never a member of the Analytical Society. He seems to have waited for the formation of the Cambridge Philosophical Society before publicly declaring his allegiance to mathematical analysis.

26. Toplis, Preface, pp. iii, iv. Toplis roundly affirmed the superiority of mathematics over other disciplines in his article in the *Philosophical Magazine* (see Note 9 above): 'Mathematics and the sciences dependent upon them, ought to make the principal part of a good education' (p. 30), and he ventured to prophesy, with some truth: 'It is possible that discoveries more wonderful and of greater utility than those already made by the help of mathematicians, may some time or other be effected, should some great genius point the way' (p. 25).
27. A.W. Thomas, p. 20. In his PhD thesis (Nottingham University, 1956), Thomas gives as his reference for this information J.A. Venn, President of Queens' College 1923–58 and son of John Venn, compiler of the *Caius Biographical History*. Attempts to trace this source for further information (1990) have proved abortive.
28. Dubbey, p. 22.
29. Hyman, p. 20.
30. Ibid., p. 24.
31. Undated letter written from London by Bromhead to Babbage, probably in 1813–14, when Bromhead was reading law at the Inner Temple. B.L. Add. MSS. 37, 182, fos. 13, 15.
32. B.L. Add. MSS. 37, 182, fo. 91.
33. Dubbey, p. 41.
34. Ibid., p. 44.
35. Gunther, p. 63. Lagrange, 'Mécanique Céleste' would appear to be a misprint: the title of Lagrange's 1788 book was 'Mécanique analytique'.
36. G.G.Cat. 4B 18.
37. Entry in D.S.B.
38. B.L. Add. MSS. 37, 182, fo. 323. Significantly, Isaac Milner was a signatory to the notice drawn up in 1819 calling the meeting at which the Cambridge Philosophical Society was founded (Twigg, p. 214).
39. The chief of these [functions] was the maintenance of the

> Philosophical Library, for there was nothing else comparable in the University, and this alone enabled Cambridge scientists to follow developments on the Continent and in the USA. As science played an even greater part in University studies, so the importance of the Library increased. (Hall, p. 28, describing changes in the University in 1865).

Chapter 4 Bromley House Library and the Essay of 1828

1. Sturges, p. 58 ff.
2. Russell, p. 7. Further information will be found in Coope and Corbett (eds), *Bromley House, 1752–1991: Four Essays Celebrating the 175th Anniversary of the Foundation of the Nottingham Subscription Library*: these cover the organization of the Library, by Peter Hoare; the history of the building, by Neville Hoskins; an account of the Library's activities, by Stephan Mastouris; and an account of the first photographic studio in Nottingham, installed on the top floor in September 1841, by Pauline F. Heathcote. Had George Green not died in May 1841, posterity might have had a photograph of him.
3. Russell, p. 18.
4. Ibid., p. 3. The Nottingham Date Book records in December 1826: 'The open sough in the Marketplace receiving the drainage from the houses in Long Row, was done way with by the substitution of an underground sewer; the Long Row itself was flagged instead of pebbled; and the market area, hitherto very unlevel and miry, was properly levelled and paved.'
5. Russell, p. 9.
6. Inkster, p. 48.
7. Ibid., p. 46.
8. The Astronomical Society received its Royal Charter in 1831. Charles Babbage was a founder member in 1820 and Sir William Herschel was the first President.
9. Wylie, commenting on the emergence of debating clubs, writes:

   > There are many intelligent people belonging to the rising classes of the town, who, unqualified by rank for an aristocratic debating society which met at Bromley House, were

still anxious to enjoy the sweets of intellectual communion with their fellowes. . . . The Nottingham Literary and Scientific Society was of a higher class than the common run of such institutions . . .' (pp. 348, 349)

The same comments would also apply to the Nottingham Subscription Library.

10. An authoritative account of the formation of the Library is given by Peter Hoare, in Coope and Corbett (eds), *Bromley House, 1752–1991*, pp. 31–5, where he describes the Library's cataloguing system. Stephan Mastouris (pp. 90–97) gives a general picture of the Library's activities.
11. H.G.G. pp. 573, 574.
12. See Feather, ch. 4, p. 44, and the section on 'Local Distribution', pp. 64–8.
13. 'Mémoire sur la Distribution de l'électricité à la surface des corps conducteurs', *Mémoires de l'Institut*, 1811, pp. 1–92. R.S.Cat. Poisson No. 25. 'Second Mémoire sur la Distribution de l'électricité à la surface des corps conducteurs', *Mémoires de l'Institut*, 1811, pp. 163–274. R.S.Cat. Poisson No. 26. 'Deux mémoires sur la théorie du magnétisme'. *Académie des Sciences*, V, 1821–2. R.S.Cat. Poisson No. 59. 'Mémoire sur la théorie du magnétisme en mouvement', *Académie des Sciences*, VI, 1823. R.S.Cat. Poisson No. 68.
14. There was one FRS in Nottingham at this time – Dr John Storer, the first President of the Library, who had been largely responsible for the foundation of the new Lunatic Asylum which was set up in Sneinton in 1812. Green's family move to Sneinton a few years later, and common membership of the Library, may have brought the elderly doctor and the younger man into contact, such that Storer was prepared to write to the President on Green's behalf. Dr Storer was not an entirely inactive Fellow of the Society. In May 1814 he sent a paper 'On an ebbing and flowing stream discovered by boring into the harbour at Bridlington', which was published in the following year's *Transactions*.
15. B.H.L. catalogue 1864 lists: *Annals of Philosophy*, 1st series, 28 vols; 3rd and 4th series, 60 vols; *Encyclopaedia Britannica*, 22 vols; *Journal of the Royal Institution*; *Metropolitan Magazine* 1831–44, 41 vols; *Revue encyclopédique*, depuis 1828, 4 vols. Dr D. Lardner, *Cabinet Cyclopaedia* (covering 63 subjects, of

which 7 cover scientific topics, the earliest being Herschel's *Discourses on Natural Philosophy* [1830]), is included in 1881 Catalogue.

16. *Edinburgh Journal of Science*, ed. Brewster, vol. 1, 1824, pp. 356–8. Ibid., vol. 5, 1826, pp. 328–30. *Quarterly Journal of Science and the Arts*, vol. 17, 1824, no. 34, pp. 317–34. Ibid., vol. 19, no. 39, 1825, pp. 122–32.
17. Essay, p. v. N.F. p.3.
18. By Dr John Roche, in a paper given at the George Green Conference held at the University of Nottingham in 1988. The question of Green's access to Poisson's memoirs is raised more acutely in Dr Roche's further comment on this passage: 'One of the key topics dealt with by Green was first introduced by Poisson in 1827.' Yet the Essay was in the press by the end of that year. 'Is it possible that Green had not seen Poisson's publication before he wrote the Essay? If so, he is far more original than we now suppose.'
19. Essay, p. viii. N.F. p. 7.
20. Essay, p. v, n. N.F. p. 4.
21. Essay, p. vi. N.F. p. 5.
22. Essay, p. vii. N.F. p. 7.
23. Green's Theorem is invariably so referred to. Reference to the functions appears to be arbitrary: 'Green' or 'Green's', 'function' or 'functions', but usually with a small 'f', except in a headline. As is known, 'Green's functions' (the most usual form) are used to help solve certain differential equations, their exact formulation depending on the demands of the problem to be solved.
24. Essay, p. vii. N.F. p. 8.
25. Two incidents may be of significant relevance to this statement. David Phillips, curator at the Castle Museum, Nottingham, and editor of *George Green, Miller, Snienton*, reported being consulted by the purchaser of a copy of a work by Cauchy bought at a second-hand bookstall in Sneinton market, and recently a Fellow of Caius College found copies of two volumes of Poisson's *Traité de mécanique* (1823) with Green's signature on the flyleaf in a second-hand bookshop in Cambridge.
26. The family of the painter Richard Parkes Bonington (1802–28) sought to repair the family fortunes by setting up a textile

business in Calais, as a result of which Bonington lived and worked in France, and became a close friend of Delacroix.

27. Green's letter to Bromhead, 19 June 1830. G.G.Cat. 4B 2.
28. H.G.G. p. 558, n. H.G. Green quotes Cajori. Florian Cajori, in *A History of Mathematics* (London: Macmillan, 1922), p. 472, states: 'About one hundred copies were printed', but produces no authority for this statement. Few copies are now extant: one sold at auction at Sotheby's in 1976 was described as 'extremely rare'.
29. Bromhead's letter of 1845. G.G. Cat. 4A 5.
30. Dr John Storer does not figure in the list of subscribers: he may, of course, have acquired a copy later.
31. R.B.N. Minute, 17 June 1830, p. 379.
32. Ibid., p. 384. A duty of the chaplain to the 'Town Jail' was to accompany condemned prisoners to the gallows. The Nottingham Date Book gives graphic accounts of the progress of the execution cart, containing criminals and clergyman. If the Reverend Samuel Lund officiated with prayers in these last moments, he showed a stronger stomach than one of his predecessors at the Free School who, after witnessing such an event, went into his class, pale and shaken, and dismissed his pupils for the day: 'Boys, I have just seen a man hanged. I cannot teach you today. You may all go home'. (A. W. Thomas, p. 140).
33. R.B.N. Minute, p. 429.
34. H.G.G. p. 590.
35. On the occasion of his visit to Nottingham in 1930. See Chapter 11, p. 159 below.

Chapter 5 Sir Edward Bromhead

1. H.G.G. p. 589.
2. For an account of the founding of Gonville Hall, see Brooke, pp. 1–12. Bromhead's paternal grandmother, Frances Gonville, was an only child and 'last of issue' (according to Simpson's *Obituaries and Records of the County of Lincoln*, 1861); she married General Bordman Bromhead in 1786. The latter's name appears on the reverse of a parchment in the possession of the Bromhead family, which bears a

Latin inscription to the effect that in the reign of Edward III, Nicholas de Gondeville was Lord of Rusheworth in Norfolk, his brother 'Edmundus' was rector of Rusheworth and Tirrington in Marshland, and this same 'Gunvil' founded an academy in Cambridge in AD 1348. The claim to 'founder's kin' carried no privileges at Caius, as it did in some other colleges.

3. *White's Directory of Lincolnshire*, 1826.
4. T.R.S. 1817.
5. Hill, *Victorian Lincoln*, p. 147.
6. Hill, *Georgian Lincoln*, p. 280.
7. Ibid., p. 287. Bayley was Sub-Dean of Lincoln Cathedral.
8. Ibid., p. 277.
9. B.L. Add. MSS. 40380 fo.56. [G.G.Cat 4B 26]
10. This letter is in private hands: G.G.Cat 4B 25. Dr John Rollett writes that the article is:

    a masterpiece, written by a man who has a total grasp of what he is expounding. Moreover he tells it in a language of the utmost simplicity, designed to convey his meaning directly to the reader, with the least possible waste of effort on the reader's part. . . . No wonder his brother was furious with him for ducking the Senate House examination, 'for everyone knew you would be Senior Wrangler'. (private correspondence)

11. Letter, 23 July 1823. B.L. Add. MSS. 37,183, fo.59.
12. Letter, 25 March 1820. B.L. Add. MSS. 37,182, fo.241.
13. For an estimate of Whewell's influence in Cambridge, see Brooke, p. 191. Brooke also recommends J.M. Clark, *Old Friends at Cambridge and Elsewhere* (London, 1900).
14. Even mature graduates were expected to wear their gowns when they visited him: Venn, p. 266, n. 1.
15. Letter, 9 June 1823. B.L. Add. MSS. 37,183, fo.37.
16. Letter, 27 December 1819. B.L. Add. MSS. 27,182, fo.184.
17. B.L. Add. MSS. 37,182, fo.196.
18. By Dr J.M. Rollett, to whom I am indebted for this information. Between 1836 and 1840 Bromhead submitted seven papers on botanical subjects: two to the *Edinburgh New Philosophical Journal*, two to the Philosophical Magazine, and three to the *Magazine of Natural History*. (R.S.Cat.)

19. *The Church of St. German, Thurlby, A Guide to Visitors*.
20. The tradition continues. Lieutenant Piers German, second son of Mr and Mrs Robin German, commanded a tank unit in the 1991 Gulf War. He named his tank Gonville Bromhead.
21. Letter, undated, B.L.Add.MSS. 37, 184, fo.330.
22. Letter, 20 April 1828. G.G.Cat. 4B 23.
23. Letter from Fisher to Bromhead, 10 May 1828. G.G.Cat.4B 24.
24. G.G.Cat. 4B 2.
25. D.P. p. 93.
26. Letter, 13 February 1830. N.U.L. MSS. G.G.Cat. 4B 3.
27. G.G.Cat. 4B 4.
28. *Pigot's Commercial Directory*, 1831.
29. There were numerous ferries across the Trent, and at Shelford the remains of the ferry stocks can still be seen. Passage by horse and even carriage was possible. Abigail Gawthern recounts in her diary how, on 9 November 1785, 'George, our postilion and the chaise horses had narrow escape from being drowned; the horses leapt out of the boat into the Trent and George was upon one of them!' (Gawthern, ed. Henstock, p. 43)
30. George Green's will. N.N.MSS. G.G.Cat. 4A 22.
31. *Noble's Gazetteer of Lincoln*, 1835.
32. G.G.Cat. 4A 5. See also Appendix IIIb below.
33. Boole was born in 1815, the son of a Lincoln cobbler; he ran an academy in Pottergate, Lincoln. He grappled unaided with Newton's *Principia* and Laplace's *Mécanique céleste*. At eighteen he gave an address on Newton when a bust was presented to the newly opened Mechanics Institute in Lincoln. Newton was a local boy – born at Woolsthorpe, Lincolnshire, in 1642, and educated at Grantham Grammar School. Bromhead was President of the Institute and presented copies of the *Philosophical Transactions* of the Royal Society to the Institute Library. By 1840 Boole was contributing to the *Cambridge Mathematical Journal* and to the Royal Society, of which he became a member in 1857. In 1849, although he had never taken a degree, he was appointed to the Chair of Mathematics in the newly established Queen's

College at Cork. He introduced what is since known as 'boolean algebra'.

34. H.G.G. p. 560. Note 21. Some later writers on Green have stated that Green sold the mill. This is an error, since he left it to his younger son. It was finally sold, derelict and ruinous, on the death of his youngest child in 1919. See Chapter 9, pp. 129–30, 134 below.

Chapter 6 The Publication of George Green's Further Investigations

1. The Bromhead correspondence was not known to H.G. Green, George Green's first biographer, writing in 1945, though he was able to see it before he died. Photographs of transcriptions of the letters are in the George Green Archive in the Manuscripts Department of Nottingham University, listed in the catalogue as follows:

4B I	19 Apr	1828	Letter from George Green to Sir Edward Bromhead
4B 2	19 Jan	1830	Letter from Green to Bromhead
4B 3	13 Feb	1830	Letter from Green to Bromhead
4B 4	17 May	1832	Letter from Green to Bromhead
4B 5	23 May	1832	Letter from Green to Bromhead
4B 6	14 Nov	1832	Letter from William Whewell to Bromhead
4B 7	21 Nov	1832	Letter from Green to Bromhead
4B 8	26 Nov	1832	Letter from Bromhead to Whewell
4B 9	18 Dec	1832	Letter from Whewell to Bromhead
4B 10	5 Jan	1833	Letter from Green to Bromhead
4B 11	n.d.	(1833)	Letter from Bromhead to Whewell
4B 12	15 Feb	1833	Letter from Green to Bromhead
4B 13	30 Mar	1833	Letter from Green to Bromhead
4B 14	1 Apr	1833	Letter from Bromhead to Whewell
4B 15	5 Apr	1833	Letter from Whewell to Bromhead
4B 16	13 Apr	1833	Letter from Green to Bromhead
4B 17	27 Apr	1833	Letter from Green to Bromhead
4B 18	28 Apr	1833	Letter from Bromhead to Whewell
4B 19	May	1833	Letter from Green to Bromhead

4B 20	1 Jun	1833	Letter from Sir David Brewster to Bromhead
4B 21	8 Jun	1833	Letter from Green to Bromhead
4B 22	22 May	1834	Letter from Green, Caius College to Bromhead

The original spelling has been retained in the passages quoted in the text.

2. G.G.Cat. 4B 4.
3. Ibid., 4B 6.
4. Ibid., 4B 14.
5. Ibid., 4B 7.
6. See Chapter 10, pp. 143–4 below.
7. G.G.Cat. 4B 10.
8. Ibid., 4B 11.
9. Ibid., 4B 3.
10. Ibid., 4B 11.
11. Ibid., 4B 17.
12. Ibid., 4B 19.
13. H.G.G. p. 568.
14. Lagrange's *Mécanique Analytique*, published in 1788.
15. Friedrich Wilhelm Bessel (1784–1846), German astronomer, friend and disciple of Gauss, who worked at Königsberg and Göttingen. He attended the meeting of the British Association for the Advancement of Science at Sheffield in 1832 in the company of Schumacher, where he met J.F. Herschel. Considered to be the founder of the German school of practical astronomy, he directed the publication of twenty-one volumes of *Beobachtungen der Königsberger Sternwarte*. Heinrich Christian Schumacher (1780–1850), astronomer and geodesist, first worked under Gauss at Göttingen; they formed a lifelong friendship. Schumacher subsequently directed the Observatory at Altona, then a Danish city. In 1821 he became an FRS and in 1823 he started the prestigious *Astronomische Nachrichten*, which is still published.

    Carl Friedrich Gauss (1777–1855), mathematical prodigy and genius, geodesist and astronomer, was the doyen of the group of German men of science of the first half of the nineteeth century which included Bessel and Schumacher, Jacobi (see below, Chapter 8, pp. 104–5), Dirichlet and Neumann. In prestige and posthumous reputation, Gauss was to Göttingen and Germany what Newton was to Cambridge and England. He evolved Gauss's Theorem, in many ways similar to Green's Theorem, the historical precedence of which was disputed – though to a lesser degree – by continental and British scientists in a manner reminiscent of the famous Leibnitz–Newton controversy. The importance of German scientists in Europe should not be forgotten. Green's knowledge of and reliance on French works, however, has inevitably focused attention on French rather than German mathematicians.

16. G.G.Cat. 4B 12.
17. Literally, 'a rough disordered pile' – a quotation from Ovid's description of Chaos in *Metamorphoses* I.7 (D.P. p. 94).
18. G.G.Cat. 4B 19.
19. Ibid. 4B 20.
20. Ibid. 4B 17.
21. The previous year (1832) Whewell had published in Cambridge *Free Motion of Points and on Universal Gravitation, including the Principal Propositions of the Principia: The First Part of a New Edition of a Treatise on dynamics*, in which he wrote:

    The really important applications of mathematics are so numerous, that it is by no means desirable to employ the student's time on detached and useless problems . . . those who would really use their mathematical requirements for their fellow students in this place, may easily find better subjects for their skill . . . I may mention as examples of such subjects, the Theory of Magnetism, as investigated by M. Poisson; Electrodynamics, according to the views of M. Ampère; the effects of Capillary Attraction as analysed by Laplace and Poisson; the Theory of Waves as treated by Poisson and others; the Theory of Tides according to the methods of Laplace. (pp. xvii–xviii)

22. Dionysius Lardner (1793–1859), educated at Trinity College Dublin, was appointed to the Chair of Natural Philosophy and Astronomy at the newly established University College London in 1827. He wrote papers for the *Edinburgh Review* and the *Encyclopaedia Metropolitana*. In 1829 he initiated, amongst other projects, the *Cabinet Cyclopaedia*, which lasted

until 1849, with 188 volumes published. He invited contributions from major contemporary figures – on the science side, from Herschel, De Morgan and Henslow.

23. G.G.Cat. 4B 21.
24. 13 April 1833. G.G.Cat. 4B 16.
25. Ibid. 4B 19.
26. Ibid. 4B 21.
27. Jane, the eldest, was ten; John, the youngest, was not yet two.
28. By Mrs Freda Wilkins-Jones, in correspondence.
29. 13 April 1833. G.G.Cat. 4B 16.
30. 17 May 1832. G.G.Cat. 4B 13.
31. 30 March 1833. G.G.Cat. 4B 13.

Chapter 7 An Undergraduate at Cambridge

1. Letter, 13 April 1833. G.G.Cat. 4B 16.
2. Bromhead's letter of 1845. G.G.Cat. 4A 5.
3. Brooke, p. 1. See also Chapter 5, Note 2 above.
4. Brooke, pp. 68, 69.
5. Ibid., p. 65
6. The asterisk denotes Fellow.
7. Venn, p. 272. John Venn entered Caius College as an undergraduate in 1853. He became a Fellow in 1857 and President in 1903; he died in 1923. In 1913 he published *Early Collegiate Life*, an account of his experience as a student. He compiled the *Biographical History of Caius College*, and also the *Alumni Cantabrigienses*, which was completed by his son, J.A. Venn. He is known outside Cambridge mainly as the inventor of 'Venn diagrams', which have made him 'a household name among the schoolchildren of the present century': Brooke, p. 233.
8. Venn, p. 270.
9. Ibid., p. 271, n.
10. Letter, 'May 1834'. G.G.Cat. 4B 22.
11. Ibid. 4B 16.
12. Ibid. 4B 21.
13. William Thomson (1824–1907), later Lord Kelvin. His biographer, S.P. Thompson, provides much information, culled from letters and diaries, about Thomson's undergraduate

life at Cambridge, where he arrived in 1841, the year after Green's departure. Thomson's papers are in the Cambridge University Library. S.P. Thompson, in his life of Lord Kelvin, quotes generously from these, so for the convenience of the general reader, references are given to his biography (S.P.T.) rather than the original papers.

14. Professor Christopher Brooke, in correspondence.
15. G.G.Cat. 4B 5.
16. Dubbey, p. 35.
17. Quoted by Ball, p. 124.
18. Notice on Green by G.J. Gray, D.N.B. 1917.
19. I am indebted to Mrs Catherine Hall, former Archivist of Caius College, for raising this point.
20. Letter, 8 June 1833. G.G.Cat. 4B 21.
21. Ibid. 4A 5. See also Appendix IIIb below. Bromhead's 'with good conduct' is rather surprising since Green was a forty-year-old undergraduate. There may be a distant echo here of a mystifying reference, in Green's letter to Bromhead of 8 June 1833, to 'the dissipations you mention' (Chapter 6, pp. 84–5 above). One wonders whether these statements are in any way relevant to the discussion on the cause of Green's death (Chapter 8, pp. 116–7 below).
22. G.G.Cat. 4B 22.
23. Lord Byron, emphatically not 'a reading man', wrote to a friend in 1807, about the time Bromhead went up to Cambridge:

    This place is wretched enough – a villainous chaos of sin and drunkenness: nothing but hazard and Burgundy, hunting, mathematics and Newmarket, riot and racing. . . . We have several parties here and this evening a large assortment of jockeys, gamblers, boxers, authors, parsons and poets, sup with me – a precious mixture but they get on well together. (*Letters of Lord Byron*, ed. R.G. Howarth, London: Dent, 1962, p. 9).

24. Quoted by Dubbey, p. 20, from G. Peacock *Observations on the Statutes of the University of Cambridge* (London, 1841).
25. S.P.T. p. 25.
26. Ibid., p. 34.
27. Ibid., p. 110. See also Ball, p. 210.
28. Dubbey, p. 28, quoting from J.M. Stair Douglas, *Life and*

Selections of correspondence of William Whewell, D.D. (London 1881), p. 56.

29. Ball, p. 193.
30. S.P.T., p. 95.
31. Hall, p. 18. William Hopkins (1793–1866) entered Peterhouse when he was thirty; he was Seventh Wrangler in 1827, FRS in 1837. He was renowned in Cambridge for his skill as a mathematics coach, and for his studies in geology. He was born in the same year as Green at Kingston-on-Soar, a village some twelve miles south of Nottingham as the crow flies. As another late entrant to Cambridge, Hopkins may later have been a friend of George Green, who passed him copies of his Essay. (See Chapter 10, pp. 143–4 below.)
32. S.P.T., p. 95.
33. Ibid., p. 103.
34. D.N.B. entry on Green. Dr Rollett has drawn my attention to the entry in *Romilly's Cambridge Diary 1832-42*, ed. J.P.T. Bury, (Cambridge: Cambridge University Press, 1967), p. 111; 'Sat. 21 Jan 1837 St John's has the first 3, viz. Griffin, Sylvester (a Jew !!!), and Brumell —— Green of Caius (son of a miller) who was expected to be SW was only 4th——'. Dr Anthony Edwards, Fellow of Caius College, found in the old College Library, as recently as March 1992, a small notebook which, through the similarity of the handwriting to that of two letters written by J.J. Smith to Bromhead in March 1845, suggests that it belonged to him. The entry for January 1837 (the year Green took the Tripos) reads: 'G. Green not Senior Wrangler – ought not to have been expected.' The entry for the following year reads: 'Jan 1838 O'Brien – as George Green in both papers.' O'Brien was Third Wrangler in 1838. This may refer to a possibility that O'Brien, like Green, was expected to be Senior Wrangler, a supposition not supported by their performance in either the first more straightforward paper or in the second more difficult problem paper.
35. N.F. p. ix.
36. William Tomlin's letter of 1845 to Dr J.C. Williams, G.G.Cat. 4A 6(ii). George Carr Peirson, born in Philadelphia, a graduate of Clare College, was Forty-fifth Wrangler in 1837.
37. Jane Smith's household, however, would probably have been more subdued than a modern family of six children. The

child mortality rate indicates that children were less robust, though Jane – unlike Green's sister, Ann Tomlin – lost no children in infancy. The eldest, Jane, at twelve, and even her younger sister, Mary Ann, at nine, would be trained to take a share in the running of the house, and care of the children. Leisure pursuits, when time permitted, had a domestic bias, if a sampler worked by 'Jane Green' at the age of ten is any indication. Of the boys, George, now seven, was of a studious disposition; John, who would die at twenty, was five. None the less, however ordered the household, there would not be much space: it is unlikely that the house in which Jane was living in Windmill Lane was any larger than 3 Notintone Place, to which Green would return in 1840.

Chapter 8 A Fellowship at Caius College

1. Electricity and magnetism as subjects of study and in-depth research at Cambridge did not come to the fore until the later nineteenth century after the advent of William Thomson and his associates; preference was given to hydrodynamics and the study of fluids, to astronomy, mechanics and optics. William Whewell had succeeded in having heat, electricity, magnetism and the wave theory of light included as examination subjects in the Mathematical Tripos in the 1830s, and had encouraged Murphy to write his book on electricity in 1833 – it being a 'traditional rule' that topics should not be introduced into the Tripos if students had no access to them in university textbooks. But in the general climate, these topics did not flourish. In 1849 the newly constituted Board of Mathematical Studies, in their report to Senate, commented on the great range of subjects covered in the examination syllabus and recommended that electricity and heat should not be included, adding that this was not an arbitrary decision but a reflection of examination practice over the years. Only one or two questions on electricity and magnetism were included each year out of a total of some 175 to 200 questions. (Ball, p. 215.)
2. N.F. p. 315.
3. N.F. p. 274. Dr G. Jaroszkiewicz has drawn my attention

to an account of Russell's subsequent papers on this subject in *Conference Proceedings: The Soliton, Introduction and Application*, ed. M. Lakshamanan (Berlin: Springer-Verlag, 1988); paper by R.K. Bullough, '"The Wave Par Excellence", the Solitary Great Wave of Equilibrium of the Fluid: An Early History of the Solitary Wave'. Russell's own paper 'Report on Waves' in the British Association Reports is dated 1844. He may have read Green's first paper on this topic in the *Transactions of the Royal Society of Edinburgh* of 1836, also his two papers of 1838 and 1839 in the *Transactions of the Cambridge Philosophical Society*. J.S. Russell was a shipbuilder who was awarded the contract to build Brunel's *Great Eastern*, launched in 1858, the largest ship ever built.

4. N.F. pp. vii–viii.
5. 'Green, though inferior to Cauchy as an analyst, was his superior in insight': Whittaker, p. 139.
6. N.F. p. ix.
7. Cross (ed. Harman), p. 215.
8. H.G.G. p. 568.
9. Whittaker, p. 153.
10. Cross (ed. Harman), p. 129.
11. See Chapter 3, Note 10, pp. 212–3.
12. These are in private hands.
13. The missing ones are 'Note on the Motion of Waves in Canals' and 'On the propagation of Light in Crystallized Media', both published in 1839.
14. Carl Jacobi (1804–51), born in Potsdam, is known for his work on elliptic functions. He taught in Königsberg for eighteen years and contributed many articles on mathematical analysis, geometry and mechanics to *Crelle's Journal*. He was one of a group – including Bessel and Franz Ernst Neumann (1798–1895) – largely centred on Königsberg, who were influential in reviving the study of mathematics in German universities. His lifelong friend, Gustave Lejeune Dirichlet (1805–59), spent many years in Paris; this helped to establish communication between French and German men of science. See Chapter 6, Note 15, pp. 223–4 for information on Gauss.
15. *Elementary Principles of the theory of Electricity, Heat and*

Molecular actions, Part I On Electricity (Cambridge: Cambridge University Press, 1833). Any other parts planned or written were never published.

16. G.G.Cat. I 6.
17. Letter, 17 May 1832, G.G.Cat. 4B 4.
18. Whittaker, p. 153.
19. H.G.G. p. 561.
20. I am indebted to Professor Christopher Brooke for this comment.
21. Green's Theorem:

    is the first example of the reciprocal relations which pervade not only Dynamics, but all branches of Physics. In the present case [the Essay], it is a relation between two different distributions of electricity, but it only needs to give suitable meanings to the symbols to translate it into the language of Hydrodynamics or Acoustics. (Lamb, p. 28)

22. Ball, p. 124.
23. S.P.T. pp. 31, 33.
24. Brooke, p. 97.
25. Ibid.
26. S.P.T. p. 109.
27. In the case of George Green, it is just remotely possible, in view of the circumstances of his departure from Caius and his death soon after (see below, pp. 115–6), and with the recent case of Murphy in mind (see below, pp. 111–2), that there were factors in Green's situation which were already causing concern.
28. H.G.G. p. 562.
29. In the case of Caius College, the statutes of 1860 abolished celibacy as a condition of holding a Fellowship. Professor Christopher Brooke is of the opinion that Caius was the first college to do so (Brooke, p. 224).
30. The term 'posted', as opposed to 'paid', signified that the wine was stored in the cellar, rather than drunk that evening (M.C. Buck, 'The Fellows' Betting Books', *The Caian*, 1982). On the subject of this particular wager, M.C. Buck reports that the Fellows adjourned to the College lawn to try out a practical experiment, and not all the eggs were broken.
31. In 1865 Isaac Todhunter published *A History of the Mathematical Theory of Probability from Pascal to Laplace*.

32. The Classical Tripos was first introduced at Cambridge in 1824, but until 1850 candidates were first required to have taken the Mathematical Tripos (Dubbey, p. 18). See also Ball, p. 38.
33. A comment made by Mrs Catherine Hall, former Archivist, Gonville and Caius College, to whom I am also grateful for the financial details concerning the Fellowships.
34. Augustus De Morgan, article on Robert Murphy in *The Penny Cyclopaedia* (1854), pp. 337, 338. This was a periodical of the Society for the Diffusion of Useful Knowledge, of which De Morgan, a friend and Cambridge contemporary of Murphy's, was a long and warm supporter, and for which he wrote several hundred articles. De Morgan's biography is listed by G.C. Smith in the bibliography of his paper on Robert Murphy.
35. This is presumably a reference to the bequest of Francis Schuldam MD who, on his death in 1776, left land to Robert Elgar, subject to the annual payment of £10 to the College for a piece of plate to be given 'to some scholar taking his degree of B.A., as after due examination they shall think most deserving' (*Caius Biographical History*). Over the years the gift had been commuted to a prize of £10, and the benefactor's name had been corrupted by various hands recording the 'Gesta'.
36. Dr Martin Davy, Master of Caius College from 1803 to 1839, was a prebendary Canon of Chichester Cathedral (Brooke, p. 211) and may have used his influence in securing Murphy's ordination in a diocese at a discreet distance from Cambridge.
37. Murphy's papers published in the Cambridge Philosophical Society *Transactions* are as follows:

    1830 On the General properties of definite integrals. *Transactions* 3 (1830), pp. 429–43. (Read May 1830.)
    (This would be the paper Whewell forwarded to Green via Bromhead: Letter dated 18 December 1832, G.G.Cat. 4B 9).

    1831 On the solution of algebraic equations. *Transactions* 4 (1833), pp. 125–54. (Read May 1831.)

    1832a On the inverse method of definite integrals with

physical applications. *Transactions* 4 (1833), pp. 353–408. (Read March 1832.) (This was the paper containing the footnote reference to Green's Essay, which alerted Thomson to its existence.)

1832b On elimination between an indefinite number of unknown quantities. *Transactions* 5 (1835), pp. 65–76. (Read November 1832.)

1833 Second memoir of the inverse method of indefinite integrals. *Transactions* 5 (1835), pp. 113–48. (Read November 1833.)

1835a Third memoir on the inverse method of indefinite integrals. *Transactions* 5 (1835), pp. 315–94. (Read March 1835.)

1835b On the resolution of equations in finite differences. *Transactions* 6 (1838), pp. 91–106. (Read November 1835.)

Other works, in addition to his book on electricity published in Cambridge in 1833 (see Note 15 above), were the *Treatise on Algebraic Equations*, published in London in 1839, two papers published in the *Transactions of the Royal Society*:

1836 First memoir on the theory of analytical operations (1837), pp. 179–210.

1837 Analysis of the roots of equations (1837), pp. 161–78;

and ten papers in the *Philosophical Magazine* (G.C. Smith, pp. 33–4).

38. De Morgan, p. 338.
39. Grattan-Guinness, 1985, p. 90.
40. Cross, p. 122.
41. The Test Act, which lifted restrictions to graduation for non-Anglicans, was repealed in 1868.
42. Comment by Bishop Harvey Goodwin, quoted in the entry on Green in D.N.B.
43. Letter of 1845, G.G.Cat. 4A 6ii.
44. Letter from the Master of St John's College, Dr R.F. Scott, to Dr E.M. Becket, G.G.Cat. 4A 16.
45. Felix Klein (1849–1925), *Vorlesung über die Entwicklung der Mathmatik der Neunzehn Jahrhundert* (Berlin, 1926), p. 14: 'Leider diente der Umstand, dass sein Talent entdeckt und aus Lichte gezogen wurde, nicht zu seinen Heil; nach

Cambridge berufen, verfiel er dem Alkohol' (translation by Mrs Jennifer Challis).

46. Archibald, p. 213.
47. A point made by Dr J.M. Rollett.
48. The gentleman in Nottingham was Dr J.C. Williams, a widely respected physician. Dr Williams contacted William Tomlin and forwarded his letter (frequently referred to in the text), together with the following note, also written by Tomlin:

    Dear Sir, I well recollect the grateful feeling which my cousin and brother in law the late Mr. George Green entertained for the kindness shown to him by Sir E.Ff. Bromhead and with this impression I think I ought to accord with his wish and yours by giving you some account of my above named relative, at the same time requesting your excuse for the imperfection of my narrative which you have herewith.

    Dr Williams forwarded both letters to Bromhead:

    April 10th 1845. Dear Sir, I have just received the enclosed letters which I forward to you – Mr. Tomlin married Mr. Green's sister, and he is a very respected man living on his Property in this town – I knew Mr.Green very well, and so far can vouch for the accuracy of his statements – I remain, Dear Sir Edward Yours truly J.C. Williams.
49. Letter from Sir Joseph Larmor, 2 October 1928. G.G.Cat. 4A 19. Certain readers of the manuscript advised against including the discussion of Green's possible alcoholism in the text. Others felt that after this lapse of time, a biography which aimed to be truthful and objective should include statements or opinions which could in some degree be substantiated. The reticence of Larmor and his generation is understandable, but the moral lapses of individuals of an earlier generation often become the recognized social problems – or in certain cases the accepted practice – of the next. Green's presumed alcoholism, the fathering of his illegitimate children, and the suicide of his son were all subjects of personal disgrace until this century, when changes in public opinion and the law have made it possible for such subjects to be discussed with candour and compassion, and without censure.

Chapter 9 George Green's Family

1. F.W.J. p. 54.
2. H.G.G. p. 564, quoting the *Nottingham Review*, 11 June 1841.
3. Mellors, p. vii. 'Principal Heaton' was Professor of Physics and Mathematics 1884–1919, and Principal of University College Nottingham 1911–29.
4. See below, pp. 126, 137; also Chapter 11, p. 160.
5. H.G.G. p. 571. The church had been rebuilt in 1837 at a cost of £4,700, and considerably enlarged to accommodate the growing population of Sneinton. The large rectory opposite was built at the same time. The church, High Anglican in character, was the first in Nottingham in 1831 to have a surpliced choir. The Liturgical Revival was started by the Reverend W.H. Wyatt, who baptized Green's last three children, Catherine, Elizabeth and Clara. The fifteenth-century choir stalls were – surprisingly – taken out of St Mary's Church, Nottingham, in 1848 and acquired for St Stephen's on payment of 10*s*. The present handsome interior is the work of Bodley in a further enlargement and redecoration completed in 1909.
6. Abigail Gawthern, in many accounts of funerals, records only two or three where the chief mourners named are women. She lists the mourners at her husband's funeral, but makes no mention of female relatives.
7. H.G.G. p. 576.
8. H.G.G. (p. 580) gives a summary of the will.
9. Certainly Belvoir Terrace, Sneinton, was built as leasehold property, four houses of which were owned by William Tomlin, according to his will.
10. F.W.J. p. 57.
11. Letter, 24 March 1919. G.G.Cat. 4A 14. This and subsequent letters between correspondents in Cambridge and Nottingham, together with Green's collected works, were presented to University College by Professor Granger's widow.
12. Abigail Gawthern's diary for 1797, p. 69:

    Mr Strey, once a grocer at the Hen Cross, died at Beeston, Apr 3, aged 86; he left most of his property to Mr. Charlton

> of Chilwell, his nephew (provided he does not marry the woman he has a large family by); if he does, he forfeits the estate.

Dr J.M. Rollett holds the view that Green honoured his father's memory by not going against his will after his death, and that this action was respected by the Church and others, who regarded Green's relationship with Jane Smith as a union blessed by God, despite the fact that they had not been formally married in church.

13. H.G.G. p. 565; on p. 566 he gives a plan of the estate.
14. F.W.J. p. 57, quoting N.R.O. M373/238 v.
15. *White's Directory*, 1858, 1862.
16. Article by Stephen Best in the *Sneinton Magazine*, no.30, Spring 1989.
17. H.G.G. p. 559. He suggests (p. 564) that 'the relationship gradually became tolerated by the Church in the circumstances in which the partners were placed', a view supported by Dr J.M. Rollett.
18. H.G.G. p. 576.
19. Letter, 12 August 1920. G.G.Cat. 4A 16. It is interesting to note that both Harvey Goodwin and Norman Ferrers were the Moderators when George Green Junior took the Mathematical Tripos, though it is perhaps unlikely that – then or later – they recognized any kinship with the mathematician.
20. Letter, 21 November 1920. G.G.Cat. 4A 18.
21. Dr J.M. Rollett, who found this advertisement, writes in correspondence:

    > I have often speculated that this was George Green Junior, having migrated to Cork by 1864 to be near George Boole, who likewise with George Green had been encouraged to continue his mathematical career by Sir Edward Bromhead. There is no record of Boole and Green having met; they were of different generations, Boole having been born in 1815. But no doubt Bromhead would have spoken about each to the other. (See Chapter 5, Note 33, p. 221–2 above).

22. Neither George Green nor his son was an MA of Cambridge.
23. See H.G.G p. 576 for this and the following details concerning the family.
24. The Hall family photograph album has three photographs,

taken in Nottingham between 1866 and the mid 1870s. Two are of Mrs Jane Moth, wearing the cameo brooch seen in all her photographs. The third is of a younger woman, probably her daughter Mary Jane Moth, since it was she who brought up her brother's children after the death of his first wife. So it seems evident that Jane continued to visit her mother in Nottingham until the latter's death in 1877.

25. H.G.G. p. 576.
26. F.W.J. p. 59.
27. Letter, 26 September 1907. G.G.Cat. 4A 10.
28. Letter, 22 March 1919. G.G.Cat. 4A 13. Finch had a particular interest in the Green family, as he explained in his letter:

    The subject [of George Green] was of much interest to me, both as a Sneinton resident and because the 3rd Wrangler in that year 1837, when Green was 4th was my uncle, Edward Brumell, who was afterwards President of St John's.

29. Letter, 24 March 1919. G.G.Cat. 4A 15.
30. The Hardwicke Marriage Act 1753 required all marriages to be solemnized in church, and duly registered. This was an attempt by the more privileged classes to prevent – or at least legalize – runaway marriages within their own ranks, and to establish a norm of sexual morality for society as a whole. It was an unpopular measure – not only because it required the payment of a fee, but because it went against custom. Different regions and occupations had their variants of 'tally weddings', when unions were solemnized through mutual declarations, the presence of witnesses, and local ritual – as, for example, at Gretna Green in Scotland. These customs persisted into the early years of the nineteenth century and were largely confined to the rural and poorer urban population. But it may be assumed that the existence of this subculture in contemporary society encouraged a tolerance of marital customs which were condemned later in the century (cf. Gillis). I am grateful to Mr C.B. Spurgin for raising this point.
31. Letter, 26 September 1907. G.G.Cat. 4A 10.
32. H.G.G. pp. 556, 593, where he quotes a letter of 19 September 1907, though there is no letter of this date in the archive.
33. H.G.G. p. 580.

34. Letter, 8 June 1833. G.G.Cat. 4A 21.
35. F.W.J. p. 60. Mrs Wilkins-Jones was the person to whom the elderly person recounted his burning of the papers. In private correspondence she gave his name as Mr Samuel Simkins, who was born and brought up in Sneinton, though he later lived in Wollaton.
36. Ibid. p. 59.
37. H.G.G. p. 567.
38. I am grateful to Mr H.B. Wilkinson of Prestwich Library, Outer Manchester, for this information, and for putting me in touch with the vicar and churchwarden of St Margaret's Church, Prestwich.
39. 'Opus sectile' is an 'Ornamental paving or wall covering from marble slabs cut in various, generally geometric shapes' (*Penguin Dictionary of Architecture*, 2nd edn, 1972). The panels, depicting Isaiah on the left and St John the Evangelist on the right, are either side of the east window, and are some ten to twelve feet high by about two feet wide.
40. Extract kindly provided by Mr P. Edwards, churchwarden of St Margaret's Church, Prestwich.
41. All these people, as well as two former servants, received bequests: 'To my nurse Henrietta Lloyd the sum of one hundred pounds . . .'. A similar sum was left to 'my coachman Edward Guest . . .' and 'my gardener Richard Mather'; a year's wages to 'each of my maid servants'; 'To my groom and two under gardeners twenty pounds each'. From the four-page will of Marion Tomlin, dated 29 August 1910.
42. Half a crown, or 2/6, was a coin representing two shillings and sixpence: there were eight half crowns to a pound (or twenty shillings). Half a crown might well have been, for example, the equivalent of a day's wage for a washerwoman at this time.
43. The reader should refer to the Green Family Tree, Appendix IVa.
44. George Green Moth was described as a telephone fitter living in Belvedere, Kent, when he sold his mother's share of the mill to Clara in 1910 (F.W.J. p. 59).
45. Any biographer of George Green would be grateful to Dr Rollett for his assiduity in researching the Moth family,

his discovery of the Bromhead correspondence, and his generosity in sharing his findings.

46. The Reform Bill, which sought to extend the franchise, was successfully passed the following year. The Duke of Newcastle, MP for Nottingham, was fanatically anti-proletarian. In 1803, instead of going to university after Eton, he went on a continental tour, after the Peace of Amiens had apparently brought the Napoleonic Wars to a close. As hostilities resumed, however, the Duke was interned in France. 'His enforced stay lasted four years, and during this period of comparative inactivity the Duke observed and encountered the egalitarianism of France after the Revolution.' This experience 'did nothing to expand his mind and everything to inspire him with a fear and hatred of liberal institutions'. The Duke became a diehard Conservative who violently opposed parliamentary reform, Catholic emancipation and Protestant dissension. 'Since his Nottingham estates included several large plots in the town as well as the Castle and Yard, Standard Hill, the Kings's Meadow and Nottingham Park, the effects of the quirks of the Fourth Duke on the town were substantial' (Brand, p. 1).

47. John Musters' wife, Mary Chaworth, and two women companions, were at home on the night of the attack, but John Musters was away. The women took refuge in the park in the pouring rain; as a result Mary Musters caught a chill and died a few days later.

48. Nottingham Castle, originally built by the Normans in the twelfth century on a motte and bailey plan, was razed by the Parliamentarians during the Civil War (1642–8). In 1661 it was bought by the Duke of Newcastle, who built a large mansion on the site, which continued to be called Nottingham Castle. This was gutted in the Reform Riots of 1831, an event featured in a drawing by Thomas Allom. In view of his hatred of the townspeople, the Duke rarely resided there, preferring his other seat at Clumber, so the Castle was not occupied on the night of the fire. The burnt out shell was bought by the Corporation of Nottingham and opened in 1878 as the first municipal museum and art gallery in the country.

Chapter 10 William Thomson and the Rediscovery of the Essay

1. William Thomson was knighted in 1866, and later created Lord Kelvin. To posterity, therefore, he is known as Kelvin; in Green's time he was known as Thomson.
2. References to these have been taken largely from S.P. Thompson's *Life of Lord Kelvin*. See Chapter 7, Note 13 above; hereafter S.P.T.
3. S.P.T. p. 83.
4. Ibid. p. 84.
5. Ibid. The apparent anomaly between the date of this entry in the notebook and the date of Thomson's letter to his father suggests that Thomson wrote the former on his journey to Paris.
6. Hugh Blackburn (1823–1909). Fifth Wrangler 1845, Fellow of Trinity 1846, admitted to the Inner Temple 1847; appointed Professor of Mathematics at Glasgow University 1849–79. Friend and fellow student of William Thomson, and student of William Hopkins.
7. S.P.T. p. 108.
8. Letter, 19 November 1907. G.G.Cat. 4A. Joseph Larmor (1857–1942) was born in Ireland and entered St John's College in 1877. He was Senior Wrangler and Second Smith's Prizeman in 1880, and was appointed Lucasian Professor of Mathematics, succeeding Stokes, in 1903; he retired in 1933. He was knighted in 1909 and received many foreign distinctions. His interest in Green led to correspondence with Kelvin and also with Professor Granger and others in Nottingham from 1907 to about 1920.
9. Compare this account with the one given on p. 143 above, written at the time of the discovery. Kelvin may be forgiven the discrepancy: his letter to Larmor was written in the year of his death, aged eighty-three.
10. Cambridge Philosophical Society *Transactions*, 4, 1833, pp. 353–408. See Chapter 8, Note 37 above for the complete list of Murphy's papers, also G.C. Smith p. 33.
11. G.G.Cat. 4B 6.
12. Kelvin, in his edition of his collected *Papers on Electrostatics*

and Magnetism (Thomson, 1st edn, 1872), states that Murphy, in the Memoir of 1833, gives a mistaken definition of the term 'potential' as used by Green. (p. 18, fn.). He carries his criticism of Murphy further in another footnote on page 2 (on the uniform motion of heat in homogeneous solid bodies and its connection with the mathematical theory of electricity):

> It is worth remarking that, referring to Green as the originator of the term, Murphy gives a mistaken definition of 'potential'. It appears highly probable that he may never have had access to Green's Essay at all, and that this is the explanation of the fact (of which any other explanation is scarcely conceivable), that in his Treatise on Electricity he makes no allusion whatever to Green's discoveries, and gives a theory in no respect pushed beyond what had been done by Poisson . . .

Murphy published his book on electricity in Cambridge in 1833, and wrote an introduction to it from Caius College, dated June 1832 – that is, some eight months before he could have read the Essay. The manuscript was probably completed before or at the same time as the introduction, and may well have been with the publishers. Murphy – presumably in straitened circumstances, as always – would not have wished, in any case, to delay publication and risk losing the market of the next academic year, so it would be unreasonable to expect any reference to the Essay in the corpus of the book.

13. An observation for which I am indebted to Dr J.M. Rollett.
14. For a detailed discussion of the differences between the Mathematics course in Cambridge and the Natural Philosophy course in Glasgow and Edinburgh, see Wilson, 'The Educational Matrix'.
15. This episode is also described in S.P. Thompson, pp. 113–19.
16. Joseph Liouville (1809–82). Charles Sturm (1803–55) was a French mathematician, born in Geneva.
17. August Leopold Crelle (1780–1855), born in Berlin. He 'had a unique sensitivity to mathematical genius. It is for this life-long unselfish intercession that Crelle deserves a place in the history of science' (Scriba, D.S.B. 1976).
18. S.P.T. pp. 91–2.
19. G.G.Cat. 4B 25, 26. Dr J.M. Rollett, who alerted me to

the existence of these two letters, was also struck by their illegibility and, at times, their incoherence.

20. Arthur Cayley (1821–95). Senior Wrangler and First Smith's Prizeman 1842, elected Fellow of Trinity the same year. Worked with James Joseph Sylvester. In 1863 he was appointed the first Sadlerian Professor of Pure Mathematics in Cambridge. Lectured at Johns Hopkins University 1881–2, at Sylvester's invitation. In 1876 he published his *Treatise on Elliptic Functions*.
21. G.G.Cat. 4A 1, 2. I am grateful to Mrs Jennifer Challis for providing this translation.
22. G.G.Cat. 4A 3. This letter in the Archive is only Thomson's draft, and as H.G. Green notes (p. 587), the list is not included in the draft.
23. See Note 9 above.
24. A.L. Crelle (ed.), *Journal fur die Reine und Angewandte Mathematik*, XXXIX (1850), pp. 73–89; XLIV (1852), pp. 356–74; XLVII (1854), pp. 161–212, Berlin. G.G.Cat. p. 5.
25. Norman Ferrers (1829–1903). Student of Harvey Goodwin. Student, lecturer and, from 1880, Master of Caius College; FRS 1887.
26. For a detailed account of the significance of the *Cambridge Mathematical Journal*, see Crosbie Smith and Norton Wise, pp. 173–6.
27. S.P.T. p. 180.
28. Ibid. p. 182; my translations.
29. Ibid. p. 183, quoting from the Report of the British Association for the Advancement of Science, 1845. Thomson was also writing of his new-found interest in Green to George Boole, who had contributed papers to the *Cambridge Mathematical Journal*. In a letter written on 11 September 1846, he asked:

    Have you seen a memoire by Green, in the Cambridge Transactions, [of 1835], 'On the Attr^{n} of Ellipsoids of variable density'? In this he gives formulae with multiple integrals (Ellipsoids of n dimension) which he applies to the case of ellipsoids by making the No. of variables 3. His results seem to be very general, but I do not know what relation in respect of generality they may bear to yours, in the Irish Transactions, which I would like to see if possible.

The next time I am at home I shall look for them in the Library. . . . The good news is that I have been unanimously elected professor of natural philosophy . . . (Sent to David Phillips at the Castle Museum, Nottingham, in 1974, by J.T. Lloyd, Department of Natural Philosophy, University of Glasgow, as part of an unpublished letter discovered in the Library Collection, *c.* 1966.)

30. S.P.T. pp. 146–9.
31. Thomson, p. 17. See also his footnote, pp. 17, 18, in which he lists subsequent articles on the theme of the Essay: these are probably the ones he included when he sent his introduction to the Essay to Crelle in 1845. He also includes a further reference to Murphy's footnote to his paper on definite integrals.
32. Ibid., p. 194, n.
33. Ibid., p. 179.
34. Ibid., p. 51.
35. E.M. Parkinson, D.S.B.
36. Stokes, p. 165.
37. Ibid., p. 168.
38. Ibid., p. 178.
39. Ibid., p. 168.
40. Ibid., p. 173.
41. Ibid., p. 178.
42. Whittaker, p. 153.
43. J. Schwinger, 'On Quantum Mechanics and the Magic Moment of the Electron', *Physical Review*, vol. 73, 30 December 1947, pp. 416–17.

    S. Tomonaga, 'On Infinite Field Functions in Quantum Field Theory', *Physical Review*, vol. 74, 1 June 1948, pp. 224–5.

    R.P. Feynman, 'Space Time Approach to Non-Relativistic Quantum Mechanics'. *Reviews of Modern Physics*, vol. 20, no. 2, April 1948, pp. 367–87.

    R.P. Feynman, 'A Relativistic Cut-off for Classical Electrodynamics', *Reviews of Modern Physics*, vol. 74, no. 8, June 1948, pp. 939–46.

    R.P. Feynman, 'Relativistic Cut-off for Quantum Electrodynamics', *Reviews of Modern Physics*, vol. 74, no. 10, July 1948, pp. 1430–38.

R.P. Feynman, 'The Theory of Positrons', *Physical Review*, vol. 76, no. 6; pp. 749–59.

F.J. Dyson, 'Radiation Theories of Tomonaga, Schwinger and Feynman', *Physical Review*, vol. 75, no. 1, 1949, pp. 486–502.

T. Matsubara, 'A New Approach to Quantum Mechanics', *Progress of Theoretical Physics*, vol. 14, no. 4, October 1855, pp. 351–77.

44. From an article by Professor L.J. Challis, to whom I am indebted for this Note, and for Notes 45–7. The aim of solid state physics is to understand the physical properties of solids – electrical, thermal, magnetic, optical, elastic, etc. – in terms of the atoms that make them up. It is now the biggest branch of physics, and as well as providing a rich field of academic study, its understanding has led to very wide applications. All our radios, TVs, recorders, computers, calculators, medical electronics, etc., depend on the properties of semiconductors, now usually in the form of a small 'chip'. Semiconductors are solids such as silicon, whose high electrical resistance can be greatly reduced by adding traces of impurity.

Another important type of solid is the superconductor. This loses *all* its electrical resistance when it is cooled to very low temperatures, typically minus 270 degrees centigrade, though this has been raised to minus 130 degrees centigrade through the recent discovery of the 'high temperature superconductors'. Very large and powerful magnets can be made from coils of superconducting wire, and these are in increasing use in hospitals for medical diagnosis (Magnetic Resonance Scanning or Imaging), and in industrial and research laboratories for chemical analysis. There are many other examples, such as the magnetic solids used in sound and video recorders, 'floppy disks', and the optical solids used for lasers. The ability to exploit the subtle properties of solids is built on the detailed understanding of these properties gained in the research laboratory.

45. Early this century, scientific enquiry was focused more and more on the nature of matter, on atoms and their constituent particles, the electrons and nuclei. It became apparent that the laws of motion established by Newton were no longer

adequate. These had lasted for more than 200 years because the errors inherent in their use to describe the objects of our everyday world moving at 'normal' speeds are insignificant. So what might be termed the 'classical' laws of Newton suffice for the study of 'classical' physics, but the laws of 'classical physics' are totally inadequate in describing the hitherto unimaginable – and unmeasurable – motion of atoms, nuclei and nuclear particles. For this we have to use quite different laws, called quantum mechanics. Virtually all modern physics based on them (chemistry and much modern technology, atomic physics, nuclear physics, solid state physics, and so on) are described by quantum mechanics. The term 'quantum' derives from the fact that the energy of a light beam is made up of 'packets' of energy called quanta (or photons). When a hot atom cools by emitting a flash of light, it emits a single quantum.

46. Total internal reflection can occur when a light beam falls on to a surface between, for example, water and air, or glass and air, provided that the light is travelling from the optically denser material (water, glass) towards the less dense material (air). In general, some light will be reflected – the surface acts as a poor mirror – and some will be transmitted ('refracted'). However, if the angle between the beam of light and the surface is less than a particular value (about 40 degrees for water to air; about 50 degrees for glass to air), no light is transmitted, and the surface becomes a perfect mirror. This is called total internal reflection, and it was first explained by Green. It is used in optical fibres, which are long fibres of highly transparent glass. A light beam is introduced at one end, and because light cannot escape through the sides, it is totally internally reflected, so all the light emerges at the other end, even if it is 100 miles away. (This neglects losses due to impurities in the glass.) The light beam can be used to carry messages much more efficiently than a telephone or radio link.

47. Elasticity is the study of how an object is deformed when a force is applied to it. One can, for example, squeeze a rubber ball in one's hand, or bend a strip of wood. A similar thing happens when heavy traffic goes over a bridge, or when a high rise building moves with the force of the wind. If the

deformations become too large, the object will collapse or break. So theories of elasticity allow us to design structures such as buildings, dams or aircraft within safe limits.

48. In Green's Memoir No. 5, 'On the Motion of Waves in a Variable Canal of Small Width and Depth', Transactions of the Cambridge Philosophical Society, vol. VI, pt IV, 1838.
49. Some twenty letters exchanged between Boole and Bromhead have recently come to light, written between 1821, when Boole was 21, and 1851 when he was 46. He borrowed books from Bromhead and sent him copies of his early papers. Bromhead died in 1851 and his eyesight had been failing for some years, which would account for his association with Boole appearing to change in character from less frequent discussion of mathematics to general social exchanges. The letters are kept in the MSS Department of Nottingham University.

Chapter 11 'Honour in His Own Country'

1. *The Miller*, 5 May 1924, noted by F.W.J. p. 42.
2. Nottingham University College became a university in 1948. An East Midlands University involving Nottingham, Derby, Leicester and Northampton had been suggested, but this was superseded when Leicester decided to have its own university.
3. For this information I am indebted to Mr Michael Brook, former Librarian, Local Studies Section, Nottingham University Library.
4. Nottingham University Science Library, courtesy of the Librarian.
5. Page 1 of the Essay is Green's introduction. The Theorem appears on page 10, so this was perhaps a misprint.
6. Information kindly provided by Dr M. Heath of the Physics Department, University of Nottingham.
7. See plate 18. Henry Piaggio was the first professor of mathematics at Nottingham University College from 1919 to 1948 and then in the newly constituted University from 1948 to 1950. He, like Professor Granger, was interested in exploring Green's life and circumstances. He was a highly regarded mathematician whose best known work is *An Elementary Treatise*

on Differential Equations and Their Applications, published in London by Bell in 1920 and various subsequent editions. In *Disturbing the Universe*, Freeman Dyson describes how, still at school at Winchester, he spent a month's Christmas vacation working 14 hours a day through Piaggio's book and its 700 problems and solutions. Dyson pp. 11–13.

8. G.G. Cat. 4A 19. Letter, 20 October 1928.
9. It has been suggested by Mr C.B. Spurgin that Green, in his study of electricity, would have known of Benjamin Franklin's experiments with kites during thunderstorms and his invention of 'lightning rods'. In 1753 Franklin reported their installation on buildings in Philadelphia, and one was affixed to St Paul's Cathedral in 1769. Their use on windmills, however, is rare. Messrs Thompsons, millwrights for 120 years, have installed only two: on Green's Mill and on the unique eight-sailer at Heckington, Lincolnshire, where lightning conductors were installed in the 1970s after the mill was struck by lightning.
10. See p. 43.
11. Each Cambridge College was originally, and in many respects still is, autonomous with individual statutes, incomes, etc. The head of the college is usually called the Master, as in Gonville and Caius where the President is one of the twelve Senior Fellows elected for a fixed term of office to oversee college administration. Queens' College, however, owes its foundation to two queens—hence the plural. The first was Margaret of Anjou, wife of Henry II, founder of King's College. Less than thirty years later, Elizabeth Woodville, wife of Edward IV, ignored Margaret's prior claim and assumed the title of foundress in statutes which established a president and twelve Fellows (Twigg. p. 7).
12. Southwell, (sometimes pronounced 'Suthall'), a small country town some twenty miles north-east of Nottingham, is the seat of the Bishop having ecclesiastical responsibility for the county of Nottinghamshire. His throne, or *cathedra*, is in the twelfth-century Minster, famous for its carved foliage in the Chapter House.
13. 'But he that giveth his mind to the law of the most High, and is occupied in the meditation thereof, will seek out the wisdom of all the ancient, and be occupied in prophecies.

He will keep the sayings of the renowned men: and where subtil parables are, he will be there also.
He will seek out the secrets of grave sentences, and be conversant in dark parables.
He will give his heart to resort early to the Lord that made him, and will pray before the most High, and will open his mouth in prayer, and make supplication for his sins.
When the great Lord will, he shall be filled with the spirit of understanding: he shall pour out wise sentences, and give thanks unto the Lord in his prayer.
He shall direct his counsel and knowledge, and in his secrets shall he meditate'.
Ecclesiasticus, Ch. 39, verses 1, 2, 3, 5, 6, 7.

14. See p. 138 and Appendix IVa.
15. Schweber p. 334, quoting the *New York Times* in 1948.
16. Martin and Glashow (see References II).
17. Private correspondence. At that period the Open University programme (designed for distance learning by mature undergraduates across the country), was only broadcast during very late or very early hours. It thus missed a more general audience and is now presumably filed away in the Open University Archives. Schwinger later published *Einstein's Legacy*, based on his script, with numerous illustrations (see References II).
18. Schweber, Section 9.22, 'A Postscript: Tomanaga, Schwinger, Feynman and Dyson', p. 575, para. 3.
19. See p. 203.
20. Edwards, A.W.F. 'More Commemorative Windows in Hall', *The Caian*, September 1993, Caius College, Cambridge.
21. See for example pp. 28–29.
22. Short Summaries of the Afternoon Papers
    Sherrington: 'Green's functions in solid state physics'.
    Green's functions are ubiquitous in solid state physics with respect to both the quantification and conceptualisation of phenomena. Typically, many-body Green's functions are used to investigate the response of a system of strongly interacting particles to an external field. One example (from many!) is the phase-locked condensation of Cooper pairs in superconductivity.
    Jones: 'Green's functions in electromagnetism'.

Although the synthesis connoted by the term 'electromagnetism' did not exist in Green's time, the tools which he invented are of fundamental importance in electromagnetic theory. Consider, for example, the Helmholtz problem which asks for solutions to $(\tilde{\nabla}^2 + k^2)u = 0$ inside or outside a region satisfying appropriate boundary conditions on its boundary. For the interior problem, finding a Green's function is in essence tantamount to solving the Dirichlet problem for the region. Finding efficient numerical methods, particularly for the exterior problem, remains a topic of current research.

Lighthill: 'Aeroacoustical uses of Green's functions etc'.

Solving the acoustical problem for jet aircraft involves the application of Green's functions to the highly non-linear problem of air flow through the jet. The analysis shows that quiet, powerful engines should be wide-bodied, high by-pass and positioned not too close to the trailing edge of a wing.

Kirchgässner: 'A trace from Green to bifurcation'.

In his 1838 paper *On the Motion of Waves in a Variable Canal of Small Width and Depth*, Green came close to using a W.K.B. approximation. Indeed, had he not neglected all non-linear terms in his analysis, he would surely have discovered the phenomenon of bifurcation. Modern ideas about dynamical systems enable Green's analysis to be satisfactorily completed.

23. See Goodwin, Irwin, 'Dirac to be given an Honoured Place Beside Newton in Westminster Abbey', *Physics Today*, November 1995, p. 74.
24. Whittaker; see p. 27.
25. See Appendix VIb.
26. Professor J. Robert Schrieffer, Mary Amanda Wood Professor of Physics, University of Pennsylvania, in a letter to Professor L.J. Challis, Nottingham University, 26 June 1974. Schrieffer shared the 1972 Nobel Prize for Physics with Bardeen and Cooper for their theory of superconductivity and is now Chief Scientist at the US National High Magnetic Field Laboratory, Tallahassee.

'. . .My first introduction to Green's functions occurred at an undergraduate mathematics course at MIT in connection with the solution of linear differential equations. While the generality of Green's work was not clear to me at that time, the simplicity and beauty of his analysis was immediately evident. In my graduate studies the quantum field theoretical applications of Green's work made a very deep impression on me. Feynman's propagators in quantum electrodynamics vividly portrayed the time evolution of the electromagnetic field and, in essence, was an application of Green's work. From that time on I have, in most of my scientific publications, dealt in one way or another with the techniques of Green functions. It is significant to note that not only are Green functions of great significance to the theoretical physicist in the solution of physical problems, these functions also are directly related to physical observations in the laboratory. Almost every experiment which weakly probes a physical system can be described in terms of the relevant Green function for this observation. Thus, the theoretical physicist has a direct link to the experimental results through the work of George Green...'

27. Sir George Dyson was an eminent musician, composer and music educationalist during the early and middle part of the twentieth century. His compositions are at present being revived with considerable interest.
28. Dyson, F.J., *Disturbing the Universe* (1979) Harper & Row, New York, pp. 84–93.
29. See above, pp. 27, 33, 212. The passage quoted is on p. 25 of Toplis's article.
30. See above, pp. 23, 24. In 1996 the vault in St. Lawrence's churchyard, South Walsham, in which John Toplis and his wife are buried, was restored. The project was led by Mr Basil Green, a fellow scientist and life-long friend of Professor Challis. Green enthusiasts lent their support and attended a ceremony in celebration in St. Lawrence's Church. Clearing away decades of weeds and brambles revealed a substantial table tomb with a slab leading to a vault. The inscription on the lid could just be deciphered and was re-inscribed.

Appendix I The Mathematics of George Green

1. The Divergence Theorem is also known as Gauss's or Ostrogradsky's Theorem. In many English textbooks published earlier in this century, it was known as Green's Theorem (see Rutherford, p. 74).

2. $$\oint_{\partial G}(Pdy - Qdx) = \iint\left(\frac{\partial P}{\partial x} + \frac{\partial Q}{\partial y}\right)dxdy.$$

   If $P(x, y) = x^3$ and $Q(x, y) = 2x + y^3$, then the left side needs four straightforward steps to arrive at the value $3a^4/2$, whilst the right-side computation takes ten steps, including manipulation of trigonometrical formulae, to reach the same value (Fraleigh, p. 660).
3. See Ferraro, p. 121.

4. N.F. gives $-\frac{1}{dr^2}$; the Essay of 1828 (p.12) gives $-\frac{1}{r^2}$.

5. Note that Green uses the word 'singularity' here not in the context of singular values, but in the sense of the equation being 'remarkable'.

6. Neumann first introduced the term Green's function in a paper printed by Halle in Leipzig in 1861.

7. The Dirichlet Principle was also named after William Thomson (1847). It is sometimes referred to as the first boundary-value problem of potential theory.
8. Bose and Joshi (p. 277) give a brief account of the method, which is termed the W.K.B. method by physicists, but is also called the Liouville-Green method by mathematicians.
9. See Jeffreys.
10. See Brillouin; Kramers; Mott; Wentzel.
11. See 'On the application of the quantum theory to atomic structure. Part I: The fundamental postulates of N. Bohr', *Proceedings of the Cambridge Philosophical Society*, Supplement, 1924; de Broglie and Brillouin; Flint; Schrödinger.
12. For a brief account of the use of a Green's tensor, see Harman, pp. 132, 133.

References I

Biographical

Aiton, E.J. (1985) *Leibnitz: A Biography*, Bristol/Boston, MA: Adam Hilger.

Amos, G.S. (1981) *A History of South Walsham*, Norfolk, private publication.

Archibald, R.C. (1945) 'James Joseph Sylvester', in Ashley Montague (ed.), *Studies and Essays in the History of Science and Learning*, New York: Schuman.

Archive Teaching Unit 2. (1974) *Working Class Unrest in Nottingham 1800–1850*, MSS. Department, Nottingham University.

Archive Teaching Unit 3. (1975) *Public Health and Housing in Early Nineteenth Century Nottingham*. MSS. Department, Nottingham University.

Artz, F.B. (1966) *Development of Technical Education in France*, Cambridge, MA: M. I. T. Press.

Babbage, C. (1830) *Reflections on the Decline of Science in England*, London: Fellowes.

Ball, W.R. (1889) *A History of Mathematics in Cambridge*, Cambridge: Cambridge University Press.

Becket, E.M. (1921) 'George Green, Mathematician 1793–1841', Thoroton Society *Transactions*, MS. copy G.G.Cat. GG3.

Blackner, J. (1815) *The History of Nottingham*, republished in facsimile by the Amethyst Press, Otley, West Yorkshire, 1985.

Brand, K. (n.d.) *The Park Estate, Nottingham*, published by the Nottingham Civic Society.

Brooke, C. (1985) *A History of Gonville and Caius College*, Woodbridge, Suffolk/Dover, NH: Boydell Press.

Cannell, D.M. (1988) *George Green, Miller and Mathematician 1793–1841*, City of Nottingham Arts Department.

Cannell, D.M. (1993) *George Green, Mathematician & Physicist, 1793–1841*, London: The Athlone Press.

Cannell, D.M. (1999) 'George Green - An Enigmatic Mathematician', *American Mathematical Monthly*, vol. 106, no. 2, February, pp. 136–151.

Cannell, D.M. and Lord, N.J. (1993) 'George Green, Mathematician and Physicist, 1793–1841', *Mathematical Gazette*, vol. 7, no. 478 (March) pp. 26–51. Published by the Mathematical Association, Leicester.

Chambers, F.D. (1945) *Modern Nottingham in the Making*, Nottingham: Nottingham Journal.

Chapman, S.D. (1962) 'The Evangelical Revival and Education', Thoroton Society *Transactions*, vol. 66, pp. 35–67.

Coope, R.T. and Corbett, J.Y. (eds.) (1991) *Bromley House, 1752–1991: Four Essays Celebrating the 175th Anniversary of the Foundation of the Nottingham Subscription Library*, Peter Hoare, Neville Hoskins, Stephan Mastouris and Pauline F. Heathcote, Nottingham Subscription Library Ltd.

Crosbie Smith and Norton Wise, M. (1989) *Energy and Empire: A Biographical Study of Lord Kelvin*, Cambridge: Cambridge University Press.

De Morgan, A. (1854) 'Robert Murphy', in the *Penny Encyclopaedia*, London: Society for the Diffusion of Useful Knowledge, vol. 2, pp. 337–338.

Dubbey, J.M. (1978) *The Mathematical Works of Charles Babbage*, Cambridge: Cambridge University Press.

Edge, D.A. (1978) 'George Green, Nottingham's Neglected Genius', *Bulletin of the Institute of Mathematics and Its Applications*, vol. 14, pp. 170–175.

Feather, J. (1985) *The Provincial Booktrade in Eighteenth Century England*, Cambridge: Cambridge University Press.

Gillis, J.R. (1987) 'Married but not Churched: Plebeian Sexual Relations and Marital Nonconformity in Eighteenth Century Britain', in R.P. Maccubbin (ed.), *Tis Nature's Fault: Unauthorized Sexuality during the Enlightenment*, Cambridge: Cambridge University Press.

Goodacre, R. (1808a) *Prospectus of Standard Hill Academy*, Nottingham County Library.

Goodacre, R. (1808b) *An Essay in the Education of Youth*, Nottingham County Library.

Grattan-Guinness, I. (1981) 'Mathematical Physics in France, 1800–1840', in J.W. Dauben (ed.), *Mathematical Perspectives: Essays on Mathematics and Its Historical Development*, New York: Academic Press, pp. 95–138.

Gray, A. (1908) *Lord Kelvin*, London: Dent.

Gunther, R.T. (1937) *Early Science in Cambridge*, Oxford: Oxford University Press.

Hall, A.R. (1969) *A History of the Cambridge Philosophical Society, 1819–1969*, Cambridge: Cambridge University Press.

Henstock, A. (ed.) (1978) *Diary of Abigail Gawthern of Nottingham 1751–1810*, Thoroton Society Record Series, vol. XXXIII.

Hill, J.W.F. (1966) *Georgian Lincoln*, Cambridge: Cambridge University Press.

Hill, J.W.F. (1972) *Victorian Lincoln*, Cambridge: Cambridge University Press.

Hoskins, W.G. (1977) *The Making of the English Landscape*, (2nd ed.), London: Hodder & Stoughton.

Hyman, A. (1974) *Charles Babbage, Pioneer of the Computer*, Oxford: Oxford University Press.

Inkster, I. (1978) 'Scientific Culture and Education in Nottingham 1800–1843', Thoroton Society *Transactions*, vol. 82, pp. 46–50.

Magliulo, B. (1982) *Les Grandes Ecoles*, Paris: Presses Universitaires de France.

Mellors, R. (1914) *Old Nottingham Suburbs Then and Now*, Nottingham: Bell.

Mitchell, J.R. *Nottingham Hospitals*, Booklet No. 4, Nottingham Civic Society.

Montague, M.A. (ed.) (1945) *Essays and Studies in the History of Science and Learning*, New York: Schuman.

Morris, C. (ed.) (1982) *The Illustrated Journeys of Celia Fiennes*, Exeter, Devon: Webb & Bower.

Field, H. (1980) *The Date Book of Nottingham* 1750–1979. Nottinghamshire County Council.

Nottingham Subscription Library (Bromley House): Catalogues and Minutes, Misc. Ms. Record Room.

W. Stevenson et al., eds., *Records of the Borough of Nottingham*, I–IX, *Nottingham*, 1882–1956, Nottingham Borough Council.

Record of the Family of Goodacre: Bromley House Library, Ms. Bound, vol. 6a/65/7.

Roche, J.J. (1988) 'Green's Contributions to Mathematical Electricity and Magnetism', paper read at the George Green Conference, Nottingham University.

Roderick, G.W. (1967) *The Emergence of a Scientific Society in England 1800–1860*, London: Macmillan.

Russell, J. (1916) *History of Nottingham Subscription Library*, private publication, printed by Derry & Sons, Nottingham.

Smith, G.C. (1984) 'Robert Murphy', *History of Mathematics*, no. 31, Mathematics Department, Monash University, Victoria, Australia.

Sturges, P. (1989) 'The Place of Libraries in the English Urban Renaissance', *Libraries and Culture*, vol. 24, no. 1, pp. 57–68.

Syer, G. (1995) 'A Visit to Nottingham in 1828', *Nottinghamshire Historian*, Nottinghamshire Local History Association.

Thomas, A.W. (1956) *A History of Nottingham High School 1573–1953*, Nottingham: Bell.

Thomas, J.M. *Michael Faraday and the Royal Institution: The Genius of Man and Place*, Bristol: Adam Hilger.

Todhunter, I. (1876) *William Whewell: An Account of His Writings, with Selections from His Literary and Scientific Correspondence* (2 vols.), London: Macmillan.

Twigg, J. (1987) *A History of Queens' College, Cambridge*, Woodbridge, Suffolk/Dover, NH: Boydell Press.

Venn, J. (1898) *A Biographical History of Gonville and Caius College, Cambridge*, Cambridge: Cambridge University Press.

Wadsworth, F.A. (1941) 'History of Two Nottingham Schools', Thoroton Society *Transactions*, vol. 45, pp. 57–75.

Wardle, S. (1971) *Education and Society in Nineteenth Century Nottingham*, Cambridge: Cambridge University Press.

Wells, R.A.E. (1983) *Riot and Political Disaffection in Nottinghamshire in the Age of Revolutions 1776–1803*, Local History Paper No. 2, Department of Adult Education, University of Nottingham.

Whittaker, E. (1910) *History of the Theories of Aether and Electricity*, London: Nelson.

Wylie, W.H. *Old and New Nottingham*, Nottingham: Nottingham Journal.

References II

Scientific*

Alexandrov, A.D., Kolmogorov, A.N. and Lavrent'ev M.A. (eds.) (1963) *Mathematics: Its Contents, Methods and Meanings*, Cambridge, MA: M.I.T. Press.

Archibald, T. (1989) 'Connectivity and Smoke Rings: Green's Second Identity in Its First Fifty Years', *Mathematics Magazine* vol. 62, pp. 219–232.

Barton, G. (1991) *Elements of Green's Functions and Propagation: Potentials, Diffusion and Waves*, Oxford: Oxford University Press.

Bose, R.K. and Joshi, M.C. (1984) *Methods of Mathematical Physics*, New York: McGraw-Hill.

Brillouin, L. (1926) 'La mecanique ondulatoire de Schrodinger; une methode generale de resolution par approximations successives.' *C. R. Acad. Sci.* vol. CLXXXIII, pp. 24–26.

Burkhardt, H. and Mayer, F.W.F. (1900) 'Potentialtheorie', *Encyklopaedie der mathematischen Wissenschaften*, vol. 2, pt. A. pp. 464–503 (article IIA7b).

Butkov, E. (1968) *Mathematical Physics*, Reading, MA: Addison-Wesley.

Cajori, F. (1962) *A History of Physics*, New York: Dover.

Cannell, D.M. and Lord, N.J. (1993) 'George Green, Mathematician and Physicist 1793–1841', *The Mathematical Gazette*, vol. 77, no. 478, (March), pp. 26–51. Published by the Leicester Mathematical Association.

Challis, L.J. (1987/88) 'George Green - Miller, Mathematician and Physicist', *Mathematical Spectrum*, vol. 20, no. 2, pp. 45–52.

*The names of scientists who gave the Royal Society lectures are not entered here; see Ch. 11, Note 22, pp. 289–290.

Collins, R.E. (1968) *Mathematical Methods for Physicists and Engineers*, New York: Rheinhold.

Coulson, C.A. (1948) *Electricity*, Edinburgh: Oliver & Boyd.

Cross, J.J. (1985) 'Integral Theorems in Cambridge Mathematical Physics 1830–1855', in P.M. Harman (ed.), *Wranglers and Physicists: Studies in Cambridge Physics in the Nineteenth Century*, Manchester: Manchester University Press, pp. 112–148.

Dyson, F.J. (1) (1979) *Disturbing the Universe*, New York: Harper & Row.

Dyson, F.J. (2) (1993) 'George Green and Physics', *Physics World*, vol. 8, 1993, pp. 33–38 (a shorter version of App. IIIb; the latter was first published in the *University of Nottingham Gazette* in 1993).

Dyson, F.J. (3) See Schweber S.S. (1994) *Q.E.D. and the Men Who Made It*, Princeton, NJ: Princeton University Press, pp. 474–575, and references therein; pp. 684–686.

Farina, J.E.G. (1976) 'The Work and Significance of George Green, the Miller Mathematician, 1793–1841', *Bulletin of the Institute of Mathematics and Its Applications*, vol. 12, no. 4, pp. 98–105.

Ferraro, V.C.A. (1956) *Electromagnetic Theory*, London: University of London.

Ferrers, N. (ed.) (1871) *Mathematical Papers of the Late George Green, Fellow of Gonville and Caius College, Cambridge*, London: Macmillan and Co. (subsequently reprinted by A. Hermann, Paris, 1903. A Facsimile of the 1871 edition was published by Chelsea Publishing Co., New York, in 1970).

Forsyth, A.R. (1890–1906) *Theory of Differential Equations*, vols. 1–6, Cambridge: Cambridge University Press.

Fraleigh, J.B. (1979) *Calculus with Analytic Geometry*, Reading, MA: Addison-Wesley.

Grattan-Guinness, I. (1981) 'Mathematical Physics in France, 1800–1840', in J.W. Dauben (ed.) *Mathematical Perspectives: Essays on Mathematics and Its Historical Development*, New York: Academic Press, pp. 95–138.

Grattan-Guinness, I. (1995) 'Why Did George Green Write His Essay of 1828?', *American Mathematical Monthly*, vol. 102, no. 5, 1995, pp. 387–396.

Gray, J. (1) (1993) 'There Was a Jolly Miller. . .' *New Scientist*, vol. 139, no. 1881, pp. 24–27.

Gray, J. (2) (1994) 'Green and Green-functions', *Mathematical Intelligencer*, vol. 16, no. 1, pp. 45–47.

Green, George (1828) *An Essay on the Application of Mathematical Analysis to the Theories of Electricity and Magnetism*, printed for the author by T. Wheelhouse, Nottingham. Subsequent publications: Crelle, A.L. (ed.) (1850–1854) *Journal für die Reine und Angewandte Mathematik*, vol. XXXIX (1850) pp. 73–89; vol. XLIV (1852) pp. 356–374; vol. XLVII (1854) pp. 161–212; Ferrers, N.M. (ed.) (1871) (see above); Ostwald, W. (1895) *Klassiker der Exacten Wissenschaften*, Leipzig: W. Engelmann (the first German translation of the *Essay*); (1958) Facsimile Edition (1,000 copies) of the *Essay*, reproduced and printed by Stig Ekelöf Göteborg, Sweden; (1995) *The Scientific Papers of George Green*, Vol. I The Essay of 1828; Vol. II Fluids and Forces; Vol. III Vibrations and Forces, George Green Memorial Fund, paperback edition.

Green, H.G. (1945) 'A Biography of George Green', in M. Ashley Montague (ed.), *Studies and Essays in the History of Science and Learning*, New York: Schuman, pp. 545–94.

Hall, G. (1998) 'George Green - Who ?', *Mathematics Today*, April 1998, pp. 48–50.

Harman, P.M. (ed.) (1985) *Wranglers and Physicists: Studies in Cambridge Physics in the Nineteenth Century*, Manchester: Manchester University Press.

Ince, E.L. (1939) *Integration of Ordinary Differential Equations*, Edinburgh: Oliver & Boyd.

Jackson, J.D. (1962) *Classical Electrodynamics*, New York: Wiley.

Jeffreys, H. (1924) 'On Certain Approximate Solutions of Linear Differential Equations of the Second Order', *Proceedings of the London Mathematical Society*, vol. 23, pp. 428–436.

Jeffreys, H. and Jeffreys, B. (1946) *Methods of Mathematical Physics*, (especially Ch. 17), reprinted 1998 in paperback, Cambridge: Cambridge University Press.

Kai Lai Chung, (1995) *Green, Brown and Probability*, River Edge, NJ: World Scientific.

Kline, M. (1972) *Mathematical Thought from Ancient to Modern Times*, Oxford: Oxford University Press.

Kramers, H. A. (1926) 'Wave Mechanics and Semi-numerical Quantisation, *Zeitschrift für Physik*, vol. XXXIX, pp. 828–840.

Love, A.E.H. (1906) *A Treatise on the Mathematical Theories of Elasticity* (2nd ed.), Cambridge: Cambridge University Press.

Lamb, H. (1895) *Hydrodynamics*, Cambridge: Cambridge University Press.

Lamb, H. (1924) *The Evolution of Mathematical Analysis*, Rouse Ball Lecture 1924, London: Cambridge University Press.

Lyuibimov, Yu. A. (1994) 'George Green: His Life and Works (on the Occasion of the Bi-centenary of His Birthday)', *Physics - Uspecki*, vol. 37, no. 1, pp. 97–109. (Ó Uspecki Fizichskikh Naukand Turpion Ltd.)

Martin, P.C. and Glashow, S.L. (1995) 'Julian Schwinger, Prodigy, Problem Solver, Pioneering Physicist', *Physics Today*, October 1995, pp. 40–46.

Maxwell, James Clerk (1892) *A Treatise on Electricity and Magnetism*, London: Clarendon Press.

Mott, N.F. (1962) *Elements of Wave Mechanics*, Cambridge: Cambridge University Press.

Neumann, C. (1862) *Allgemeine Lösung des Problemes über den stationärer Temperaturzustand eines homogenen Körpers welcher von irgend zwei nichtconcentrischen Kugelflächen begrenzt wird*, Leipzig: Halle, (the first named reference to Green's functions).

Piaggio, H.T. (1920) *Differential Equations*, London: Bell.

Rickayzen, G. (1980) *Green's Functions and Condensed Matter*, London: Academic Press.

Rickayzen, G. (1965) *Theory of Superconductivity*, New York: Wiley-Interscience.

Roach, G.F. (1982) *Green's Functions*, Cambridge: Cambridge University Press.

Roche, J. (1988) 'Green's Contribution to Mathematical Electricity and Magnetism', Paper read at George Green Conference, Nottingham University.

Rutherford, D.E. (1946) *Vector Methods*, 4th ed., Edinburgh: Oliver & Boyd.

Schrödinger, E. (1928) *Collected Papers on Wave Mechanics*, translated by J.F. Shearer and W.M. Deans, London: Blackie.

Schweber, S.S. (1994) *Q.E.D. and the Men Who Made It*, Princeton, NJ: Princeton University Press.

Schwinger, J. (1) (1986) *Einstein's Legacy: The Unity of Space and Time*, New York: Scientific American Books Inc.

Schwinger, J. (2) See Schweber S.S. (1994) *Q.E.D. and the Men Who Made It*, Princeton, NJ: Princeton University Press, pp. 273–372; 572–575, references therein; pp. 717–718.

Steadman, G.E. (1968) 'Green's Functions', *Contemporary Physics*, vol. 1, pp. 49–69.

Stokes, G.G. (1880) *Mathematics and Physics Papers*, Cambridge: Cambridge University Press.

Stolze, C.H. (1978) 'The Divergence Theorem', *Historia Mathematica*, vol. 5, pp. 437–442.

Temple, G. (1981) *A Hundred Years of Mathematics*, London: Duckworth.

Thomson, W. (1884) *Papers on Electrostatics and Magnetism* (2nd ed.), London: Macmillan. (1st ed. 1872).

Wentzel, G. (1926) 'A generalisation of the quantum conditions for the purposes of wave mechanics', *Zeitschrift für Physik*, vol. XXXVIII, pp. 518–529.

Wallis, P.J. (1978) 'George Green', in Gillispie, C.C., ed., *Dictionary of Scientific Biography*, Supplement 1, vol. XV, Schribner: New York, pp. 199–201.

Whitrow, G.J. (1984) 'George Green (1793–1841): A Pioneer of Modern Mathematical Physics and Its Methodology', *Annali dell' Instituto e Museo di Storia della Scienza di Firenze*, anno 9, facicolo 2.

Wilson, D.B. (1985) 'The educational matrix: Physics education at early-Victorian Cambridge, Edinburgh and Glasgow Universities', in P.M. Harman (ed.), *Wranglers and Physicists: Studies on Cambridge Physics in the Nineteenth Century*, Manchester: Manchester University Press, pp. 12–48.

Whittaker, E. (1910) *History of the Theories of Aether and Electricity*, London: Nelson.

Zauderer, E. (1989) *Partial Differential Equations of Mathematics*, New York: Wiley.

Websites:

1. *George Green* (George Green Memorial Fund
c/o Physics Department, Nottingham University, U.K.)
http://www.nottingham.ac.uk/physics/green/homepage.htm
email: linda.wightman@nottingham.ac.uk

2. *George Green* (Nottingham University
Department of Manuscripts and Special Collections)
http://www.nottingham.ac.uk/library
email: dorothy.johnston@nottingham.ac.uk

3. *Green's Mill* (Nottingham, U.K.)
http://www.innotts.co.uk/greensmill

4. International Astronomical Union (I.A.U.),
Gazetteer of Planetary Nomenclature
(for details of the George Green crater on the moon)
http://wwwflag:wr.usgs.gov/USGSFlag/Space/nomen/nomen.html
email: kaare.aksnes@astro.uio.no

Index